Autonomous Vehicles Volume 1

Scrivener Publishing
100 Cummings Center, Suite 541J
Beverly, MA 01915-6106

Publishers at Scrivener
Martin Scrivener (martin@scrivenerpublishing.com)
Phillip Carmical (pcarmical@scrivenerpublishing.com)

Autonomous Vehicles Volume 1

Using Machine Intelligence

Edited by
Romil Rawat
A. Mary Sowjanya
Syed Imran Patel
Varshali Jaiswal
Imran Khan
and
Allam Balaram

Scrivener Publishing

WILEY

This edition first published 2023 by John Wiley & Sons, Inc., 111 River Street, Hoboken, NJ 07030, USA and Scrivener Publishing LLC, 100 Cummings Center, Suite 541J, Beverly, MA 01915, USA

For more information about Scrivener publications please visit www.scrivenerpublishing.com.

Wiley Global Headquarters
111 River Street, Hoboken, NJ 07030, USA

For details of our global editorial offices, customer services, and more information about Wiley products visit us at www.wiley.com.

Library of Congress Cataloging-in-Publication Data

ISBN 9781119871958

Cover image: Pixabay.com
Cover design by Russell Richardson

Set in size of 11pt and Minion Pro by Manila Typesetting Company, Makati, Philippines

Printed in the USA

Contents

Preface

This groundbreaking new volume provides and addresses the current challenges, approaches, and applications relating to autonomous vehicles using Internet-of-Things (IoT), Machine Learning (ML), Deep Learning, Data Science, Cloud, Fog and Edge Computing, Computer Vision, ANN, and Artificial Intelligence (AI) techniques and covers the applications in driverless intelligent systems based on advance computer science algorithms applicable for business engineering, automation, heath informatics, military applications, transportation, education, and so on. Several companies are working on autonomous systems and developing cars, truck, drones and even aero-taxis (unmanned aerial vehicles [UAVs]) for frequent travelling having an easy, flexible, and fast mode of transportation on the road or in the air for sharing real time data and activities. Several self-driving or autonomous ("driverless") cars, trucks, and drones incorporate a variety of IoT devices and sensing technologies such as sensors, gyroscopes, cloud computing, and fog layer, allowing the vehicles to sense, process, and maintain massive amounts of data on traffic, routes, suitable time to travel, potholes, sharp turns, and robots for pipe inspection in the construction and mining industries. Autonomous vehicle is a broad class of automation used for farming vehicles such as irrigators, tractors, and buggies, mining vehicles such as drilling rigs, with industrial vehicles like forklifts and car crash testing vehicles.

The Autonomous Vehicles project is a drawn-out logical examination program that plans to contemplate and comprehend these worldwide wonders through a computational, data-driven methodology. We have created different multilingual data mining, text mining, and web mining procedures to perform connection examination, content investigation, web measurements (specialized refinement) examination, notion investigation, origin investigation, and video investigation in our exploration. The methodologies and strategies created in this venture add to propelling the field of Autonomous Vehicles applications.

Autonomous Vehicles: Using Machine Intelligence is organized into divergent distinct sections that provide comprehensive coverage of important topics. The sections are:

- Activity and Threat Analysis Using Machine Intelligence
- Algorithms for Autonomous Cars and Vehicles
- Drone Technology for Internet of Vehicles
- IoT-Based Smart City Automation for Connected Vehicles
- Traffic Management Systems for UAV's and Smart City Models

Objectives, Impact, and Value

This book aims to convey an overview of autonomous vehicle methodologies, suggest a systematic, computational approach to handling problem-solving, and lay out various tactics and strategies. Researchers, security professionals, experts in autonomy, and strategists will all benefit from the knowledge it carries. The monograph can likewise fill in as a source of perspective material or reading material in alumni level courses identified with state of the art models of *Autonomous Vehicles: Using Machine Intelligence.*

The book is targeted to academic researchers, research scholars, practitioners, students, traffic engineers, city planners, consultants, traffic planners, vehicle designers and all key stakeholders working on intelligent engineering solutions. Contributions are based on conceptual and theoretical solutions, quantitative or empirical results, and model/simulation-based experimental studies.

1

Anomalous Activity Detection Using Deep Learning Techniques in Autonomous Vehicles

Amit Juyal[1,2], Sachin Sharma[1] and Priya Matta[1]*

[1]Department of Computer Science and Engineering, Graphic Era (Deemed to be University), Dehradun, Uttarakhand, India
[2]School of Computing, Graphic Era Hill University, Dehradun, Uttarakhand, India

Abstract

Autonomous driving is self-driving without the intervention of a human driver. A self-driving autonomous vehicle is designed with the help of high-technology sensors that can sense the traffic and traffic signals in the surroundings and move accordingly. It becomes necessary for a self-driving vehicle to take a right decision at the right time in an uncertain traffic environment. Any unusual anomalous activity or unexpected obstacle that could not be detected by an autonomous vehicle can lead to a road accident. For decision making in autonomous vehicles, very precisely designed and optimized programming software are developed and intensively trained to install in vehicle's computer system. But in spite of these trained software some of the anomalous activity could become a hindrance to detect promptly during self-driving. Therefore, automatic detection and recognition of anomalies in autonomous vehicles is critical to a safe drive. In this chapter we discuss and propos deep learning method for anonymous activity detection of other vehicles that can be danger for safe driving in an autonomous vehicle. The present chapter focuses on various conditions and possible anomalies that should be known to handle while developing software for autonomous vehicles using deep learning models. A variety of deep learning models were tested to detect abnormalities, and we discovered that deep learning models can detect anomalies in real time. We have also observed that incremental development in YOLO (You Only Look Once) make it more accurate and agile in object detection. We suggest that anomalies should be detected in real time and YOLO can play a vital role in anomalous activity.

**Corresponding author*: mattapriya21@gmail.com

Romil Rawat, A. Mary Sowjanya, Syed Imran Patel, Varshali Jaiswal, Imran Khan and Allam Balaram (eds.) Autonomous Vehicles Volume 1: Using Machine Intelligence, (1–26) © 2023 Scrivener Publishing LLC

Keywords: Autonomous self-driving, AI, deep learning, YOLO, R-CNN, Fast R-CNN, Faster R-CNN, SSD

1.1 Introduction

A crucial problem for the success of autonomous vehicles is ensuring safe driving. Before being released to the general public, self-driving cars must be thoroughly trained and tested. It should not compromise the safety of passengers or other traffic objects like vehicles, bikers, cyclists, pedestrians, etc. It should be thoroughly tested before the actual launch. Self-driving cars are controlled by software and the software must be trained in such a way that it can perform well under all circumstances or conditions. The following points need to be considered while developing software for autonomous vehicles.

Infrastructure: In the case of self-driving vehicles, infrastructure can be crucial. Almost the majority of the world's roads and transportation infrastructure are now designed for human use. Autonomous vehicles will be required to operate inside existing infrastructure. For a self-driving vehicle, it is a challenging task to use current infrastructure. The software should be trained in such a way that it can easily adapt to the existing road infrastructure.

Traffic conditions: In real time, it is very difficult to predict traffic that what will happen next. It is almost impossible to accommodate all scenarios of traffic conditions while developing software for autonomous vehicles. However, AI-based algorithms should be developed in such a way that it can learn by itself with time and experience.

Weather condition: Weather can affect driving ability in autonomous vehicles. It may be possible that the inputs from various sensors and cameras get damaged due to bad weather, and in heavy rain or in a snowstorm various road streaks and lanes information can be hidden. An autonomous vehicle navigation system should be developed while considering weather conditions and it should be trained and tested in all weather conditions.

Software security: Self-driving cars completely depend on software, and software can be hacked and can be infected by viruses (a malicious computer code). Computer viruses can cause unexpected glitches in self-driving cars. These glitches can be harmful to self-driving cars especially while driving at a high speed. So the software needs to be secure for unauthorized access and viruses for safe driving.

1.1.1 Organization of Chapter

The rest of the chapter is outlined as follows. Section 1.2 gives the literature review. Section 1.3 describes an artificial intelligence approach in autonomous vehicles, section 1.4 outlines technologies inside an autonomous vehicle, section 1.5 shows major tasks in autonomous vehicle using AI, section 1.6 shows the benefits of autonomous vehicle, section 1.7 describes applications of autonomous vehicles. In section 1.8, anomalous activities and their categorization are described, while section 1.9 describes deep learning methods in an autonomous vehicle. Section 1.10 shows the working of YOLO, and section 1.11 shows the proposed method. Section 1.12 shows the proposed algorithm, while section 1.13 is a comparative study and discussion, and section 1.14 presents the conclusion of this chapter.

1.2 Literature Review

A security model has been suggested that can deal with three types of cyber-attacks for Electronic Control Units (ECUs). Over the years, the automobile sector has improved technology and there has been advancement in car production. To make the vehicle more comfortable and automated, companies are doing research on new technology. One of the advancements is that companies are replacing some mechanical parts with electronic components to introduce automation into vehicles. ECU is an electronic control unit that can communicate with other ECUs by messages. ECUs are modern technology that relies upon Control Area Network (CAN) and ensures that all the critical parts of a vehicle, like braking, engine, airbag, steering wheel, fuel indication, and acceleration are working properly. Due to the lack of security on the CAN bus network, it can be hacked and attackers can perform malicious activities in the ECU. The author's security mechanism can solve three types of message attacks like fuzzy, Denial of service (DoS), and impersonation attacks. Deep learning-based network, Deep Denoising Autoencoder was adopted in the proposed security framework. Ecogeography-based optimization (EBO) algorithm was integrated with deep denoising autoencoder. For experimental data, malicious messages injected in CAN traffic were used. The experiment result showed that the proposed deep denoising autoencoder method outperforms the other machine learning models on three different CAN traffic datasets by achieving the highest hit rate and lowest miss rate [1].

A network called mIoUNe for detecting failure cases in semantic segmentation was proposed. To identify identical pixels and then label identical pixels with corresponding class is image segmentation. It may be possible that in the predicted semantic segmentation map, some pixels are labeled with the wrong class. For real-time applications like autonomous vehicles, this type of anomaly can lead to unsafe driving and results in accidents. The authors proposed a method using a neural network. Their network predicts the mean of the intersection of union (mIoU) to ensure that all the pixels were accurately classified. CNN and FCN were used in mIoUNet. Experimental results revealed that the proposed method achieved an accuracy of 93.21% mIoU prediction and 84.8% failure detection. In another experiment with HMG's SVM camera acquisition dataset, the method achieved 90.51% mIoU prediction accuracy and 83.33% failure detection accuracy [2].

Road infrastructure will play a key role in the success of self-driving cars. There are many causes of traffic accidents and one of them is bad conditions. Android application was developed using OpenCV library to detect potholes and cracks in roads in real time. The proposed Automatic Pavement Distress Recognition (APDR) system was developed by combining the Android framework with the Open CV library. The system can detect road anomalies like fatigue cracks, longitudinal, potholes and transversal cracks. The Local Binary Pattern (LBP) feature cascade classifier was employed to train the system for positive samples and negative samples. A custom image dataset of the streets of Rome (Italy) was constructed for the experimental work. Using the LBP feature cascade classifier, the proposed Android system can detect road anomalies directly from live video frames. The system was tested on three android devices. Results showed that the system performed well in an older version android device as well as in a new device. This showed the portability of the proposed system [3].

Connected and automated vehicles (CAVs) capture the surrounding information from various sensors and cameras. Accurate information is very important for self-driving cars because autonomous vehicles are controlled by software. It may be possible that a sensor can provide an anomalous reading due to a faulty sensor or cyber-attacks. A faulty reading in an autonomous vehicle can leads to accidents. Therefore, real-time detection of anomalies is important. The experimental result of an anomaly detection method using CNN and Kalman filtering showed that the proposed approach can detect anomalies and identify their sources with high accuracy, sensitivity, and F1 score [4].

In autonomous vehicles, LiDAR, RADAR, cameras, and various sensors work together to collect surrounding traffic information. Self-driving cars or autonomous vehicles totally depend upon electronic devices for decision-making. It may be possible that electronic devices in autonomous vehicles may fail. If an electronic component fails it can be dangerous for driving in an autonomous vehicle. A federated sensor data fusion architecture to deal with sensor faults was proposed. Their architecture can detect a faulty sensor. SVM was used to detect a faulty sensor while obstacle detection in autonomous vehicles. For experimental work, they used KITTI dataset, and results showed that the proposed architecture can detect soft and hard faults from a particular sensor [5].

In an autonomous vehicle environment, vehicle identification is a challenging task. The impeded vehicle on the road can block traffic and the result may be a traffic jam. Such types of roadside anomalies should be detected on time so that roadside assistance can be provided to the stalled vehicle. A deep learning-based method to detect anomalous vehicles and reidentification of vehicles was proposed. For vehicle reidentification, the input image was cropped in the image pre-processing step of the proposed method. In the next step, features were extracted using adaptive attention. The final step is post-processing in which relatively discriminative features were generated. Anomalous vehicles move at a slower speed than normal vehicles. Mask RCNN was used as a detector. Target vehicles are immobile and for anomalous vehicle detection, they are considered vehicles of interest. The authors proposed a two-step method for anomalous vehicle detection. In the first stage, the model was trained to identify slower vehicles than the normal vehicles by searching sequential bounding boxes that have overlapped spatially. In the second stage, the initial proposals were reined in by the establishment of a cuboid search radius over the initial search [6].

Highway traffic anomaly (HTA) dataset to detect anomalous traffic incidents and patterns was proposed. HTA dataset was systematically constructed from the Berkeley Deep Drive dataset. Berkeley Deep Drive dataset has high-resolution dash cam videos. CGAN, FlowNet, PredNet N+1, and PredNEt N +6 models were evaluated using the HTA dataset. To measure the accuracy, AUC scores were considered. The experimental result showed that state-of-the-art models achieved low AUC scores which are not acceptable in real-time applications like in self-driving cars. Except for impeded vehicles detection, PredNet N+ 6 achieved the highest AUC score among other models in detecting anomalies [7].

A deep learning-based fault detection method was proposed. Autonomous vehicles may malfunction due to electrical or mechanical failure. For safe driving in autonomous vehicles, faults must be automatically detected so that appropriate action can be taken by self-driving cars. Simulation data was generated. Using wavelet transform, the generated one-dimensional data was converted into a two-dimensional image. Three different channels of these 2D signals were fed into the deep learning model to classify whether the signal is faulty or not [8].

For safe driving in autonomous vehicles, the intelligent traffic monitoring system can be helpful for autonomous vehicles. Intelligent traffic monitoring systems can prevent traffic jams and other accidents by providing information of the time and place of accidents to other vehicles. A decision tree-based method extracts road accident anomalies from video. Deep learning (YOLOv5) was used in the proposed method. Traffic videos were sorted using a video sorting system. Then foreground object detection and background estimation were performed concurrently. Background images were fed to an object detector to characterize potential anomalies. Then a decision tree algorithm based on a predefined rule was used to classify anomalies. The proposed method achieved a 0.8571 F1 scores and 0.5686 S4 score [9].

One of the studies [10] proposed a novel unsupervised learning model based on deep autoencoder to detect the self-reported location anomaly in Connected and Autonomous vehicles (CAV) by vehicle location and the Received Signal Strength Indicator (RSSI) as features. The proposed model is trained and tested on simulation data generated by OMNeT++, modified INET framework and SUMO traffic generator, in an area of Bristol, UK. The DAE approach proposed in the study is useful in detecting CAV location anomaly in a complicated scenario. And this model connects unsupervised learning and anomaly detection in CAVs which might be beneficial as powerful properties of unsupervised learning in ITSs. During past years many researchers introduced intrusion detection methods for CAVs using deep learning and discriminated analysis [11, 12] and a CAV misbehavior detection method was also proposed for service management [13]. Usually synthetic datasets, generated by means of simulation frameworks was used for this kind of researches [14].

A study proposed Hidden Markov Model (HMM) based behavior analysis method to assess vehicles' driving behavior on the road and detect anomalous patterns. The real-time velocity and position of the surrounding vehicles-based algorithm was used, which has been provided by the Conditional Monte Carlo Dense Occupancy Tracker (CMCDOT) framework. The vehicle has been classified into several observation states, in association with road information. These were named as Approaching, Braking, Lane Changing, and Lane Keeping. By this, the ego-vehicle could observe the movements of nearby vehicles and infer their driving behavior by tracking and analyzing their motion and position. CARLA simulator was used to generate several typical abnormal movements for two vehicles. The results showed that the proposed method can successfully detect the risk moments [15].

For detection of anomalous activity and safe driving in an autonomous vehicle, approaches have focused on an anomaly segmentation framework that combines two complementary approaches to anomaly detection: uncertainty and resynthesis methods to resynthesize the image from the semantic label map to find dissimilarities with the input image. The work comprises two methodologies to produce robust predictions for anomaly segmentation [16]. A pixel-wise anomaly detection Framework for complex driving scenes (i.e., urban landscapes) were designed that uses uncertainty maps, like softmax entropy, softmax distance and perceptual differences to improve over existing resynthesis methods in finding dissimilarities between the input and generated images. This method also works for lighter segmentation and synthesis networks, making it ready for deployment in autonomous machines. It is the best overall method across datasets on the Fishyscapes benchmark and can be used with any already trained state-of-the-art segmentation model [16].

To avoid anomalies like signal interference, software and hardware errors, cyber-attacks and many more, a deep learning method, to propose a model based on robust real-time technique was approached. A LSTM auto recorder was used to extract signal features and classification. The model impact was detected using channel boosting, and accuracy obtained was 95.5% and precision of 94.2% [17].

In a real-world scenario, for safe driving, autonomous vehicles need to track another vehicle's motion also. Sudden braking by another vehicle may cause accidents in autonomous vehicles. One way to keep track of another vehicle's motion is to detect a brake light. The brake light is an indicator for other vehicles that the vehicle is about to slow down. Therefore, in self-driving cars intelligent brake light detection and recognition can help self-driving cars for flawless and secure driving. A two-stage method has been proposed to detect vehicles and identify the vehicle's tail brake light. The method is able to detect vehicles and recognize vehicle brake lights in real time from a single image. BVLC AlexNet, an 8-layer CNN model was trained on the rear vehicle database. In-vehicle detection algorithm road segmentation and vanishing point detection improves the accuracy of the proposed method. At the detection level, the lidar and vision were fused. This two-stage method is capable to detect and recognize vehicle tail brake lights from the noisy images also [18].

1.3 Artificial Intelligence in Autonomous Vehicles

Although many industries and researchers are working on autonomous vehicles, this technology still needs a lot of improvement before it really matures. Autonomous vehicle means a vehicle that can be used for transport in the real world without human intervention. But still autonomous vehicles, which were launched by many automobile industries, were tested in controlled environments. The autonomous vehicle needs a lot of improvement before it can actually face real-world conditions. In the actual world, distance from other vehicles, pedestrians, animals, speed-breakers, traffic signals, and other unpredictable dynamic surroundings all play a role in driving a safe vehicle. The autonomous car must be equipped with a reliable and sophisticated system that can correctly interpret information from numerous sensors and gadgets in order to comprehend what is going on around it and drive accordingly. An autonomous vehicle is a vehicle that can drive itself without any human intervention. Figure 1.1 shows the types of vehicles. In Figure 1.1, type VI represents the "fully autonomous" scenario whereas type I represents "fully human driving". From type I to type VI represents the levels of human and machine driving.

Many industries like agriculture, textiles, health, and automobile are already using Artificial Intelligence (AI) to automate things. The automobile industry also uses AI in vehicle manufacturing and exploring AI to build autonomous vehicles. AI can make it possible to see self-driving cars

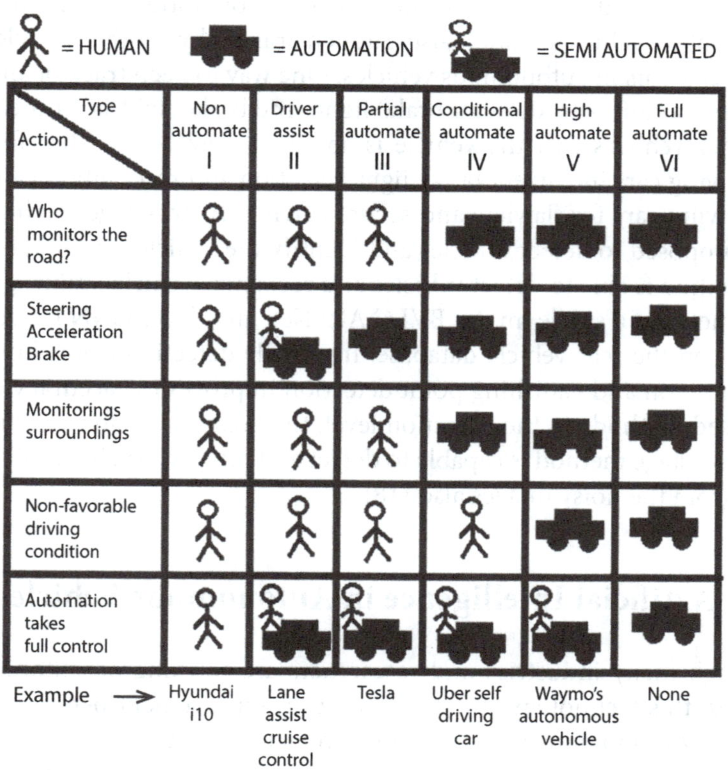

Figure 1.1 Levels of automation in vehicles.

or autonomous vehicles for everyday transportation. AI is an important and revolutionary component in manufacturing of autonomous vehicles. To mimic human driving skills is a challenging and tough task for autonomous vehicles. Autonomous vehicles rely on various sensors, cameras, and other electronic equipment to simultaneously collect surrounding traffic information. An autonomous vehicle needs intelligent software that can process the collected data and plan the trajectory for a safe autonomous drive. AI techniques such as deep learning and machine learning can be used to recognize surrounding objects, detect lanes, plan routes, and control mechanical components of a vehicle. For a safe drive-in autonomous vehicle, object detection is critical. Object detection can be accomplished using a variety of deep learning techniques such as R-CNN, Fast R-CNN, and Faster R-CNN; however, autonomous vehicles must identify barriers and objects in real time. Therefore, autonomous or self-driving cars require such algorithms that should have agility in object detection. You Only Look Once (YOLO) and Single-shot detector (SSD) are deep learning algorithms that can detect objects in real time; therefore, these algorithms can be used in autonomous vehicles. The deep learning algorithms need to interact with various hardware components like sensors, cameras for nearby data and to process data in real time. To process data in real time needs intensive computational power. Therefore, traditional CPU with limited memory is not sufficient to execute deep learning algorithms. Therefore, in autonomous vehicle extra processing power like Graphical processing unit (GPUs) for real-time object detection is required.

1.4 Technologies Inside Autonomous Vehicle

Hardware and software are the main components in autonomous vehicles that are primarily responsible for driving. Hardware (cameras, sensors, LiDAR and RADAR) is responsible for collecting data from the surroundings and software is responsible for processing the data. Computer vision, machine learning, LiDAR, and RADAR are essential technologies for autonomous vehicle functioning. In an autonomous vehicle, machine learning algorithms continuously learn for themselves from the surrounding environment and predict possible outcomes.

A. HARDWARE TECHNOLOGIES

Cameras for Computer Vision: The camera is the main source to record visual data or live video in an autonomous vehicle. This captured data is used by deep learning algorithms to detect the objects.

Radio Detection and Ranging (RADAR): RADAR technology is used to detect objects such as aircraft, ships, spacecraft, etc. It uses radio waves to determine the range, velocity and angle of objects. In an autonomous vehicle, radar is used to detect other vehicles, pedestrians, and obstacles. Radar can be used in inclement weather and low intensity of light.

Light Detection and Ranging (LiDAR): LiDAR is a remote sensing technology that uses laser light pulses to measure the distance from objects like pedestrians, other vehicles, and unmovable obstacles. To detect small objects LiDAR is better than RADAR and can build an exact 3D monochromatic image of an object. In bad weather and night-time LiDAR does not perform well, and also it is more expensive than RADAR.

Sensors: A sensor is a small device that reacts to environmental things such as light, speed, pressure, humidity, temperature. The autonomous vehicle uses sensors to detect what is actually happening around it. It also uses the traditional Global Positioning System (GPS) for path planning.

Computer resources: A fully autonomous vehicle needs extreme computational power. Graphical Processing Unit (GPU) is required with traditional Central Processing Unit (CPU) to execute and process complex Machine learning and deep learning algorithms.

B. SOFTWARE TECHNOLOGIES

In autonomous vehicles, machine learning algorithms continuously interpret the inputs from various devices and predict possible changes. As shown in Figure 1.2, the image captured by the camera goes to the processing unit for important feature selection for object detection. Depending on the features, the vehicle performs possible actions based on the detected object.

Images captured by cameras may contain irrelevant data. Pattern recognition algorithms like Support Vector Machine (SVM), K-nearest neighbour,

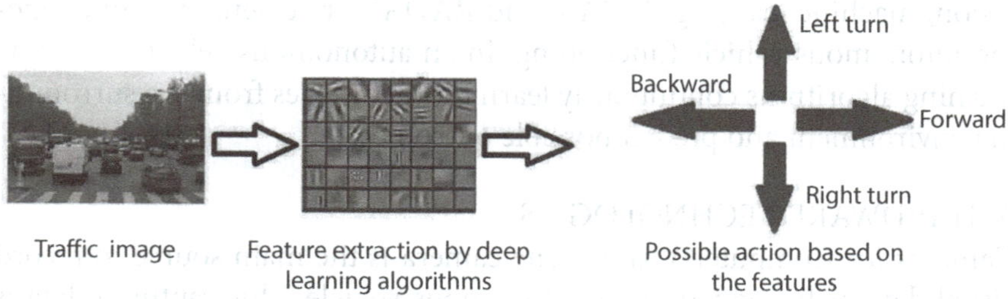

Figure 1.2 Possible vehicle action prediction based on features.

and Decision Tree are used to remove irrelevant data points. Pattern recognition helps in object classifications or object recognition.

Regression Algorithms: Regression algorithms are used for object detection and localization. Neural network regression, Random Forest and Bayesian regression are the well-known examples of regression algorithms.

Clustering algorithms: Clustering algorithms are mainly used for object recognition. It may be possible that classification algorithms may predict false results in object recognition; the reason may be deformed images, low-resolution images or blurred images. Clustering algorithms create clusters or groups of similar data points. Commonly used clustering algorithms are K-means and Multi-class neural network.

Decision-Making Algorithms: After detecting and classification of object, some algorithms are used in decision making and take appropriate action for a safe drive. Commonly used decision-making algorithms are Decision Tree and Random Forest.

1.5 Major Tasks in Autonomous Vehicle Using AI

Object classification, localization and detection are computer vision problems. Object classification is to identify the class of object present in the digital image. To find the exact position in the image is object localization and to classify and localize more than one object in a single image is object detection. Studies done in the recent past show that deep learning algorithms achieve marvelous performance in object classification, localization and detection.

Object detection is very important for safe driving in autonomous vehicles. After detection of an object, a vehicle can decide the next action, whether to take a left, a right, accelerate or brake. By detecting objects in real time, deep learning algorithms can make possible self-driving for everyday transportation in the future.

Path plan and execution: Planning better and secure paths is important in autonomous vehicles. Accidents can be avoided with accurate path planning and can be helpful in reducing traffic congestion. Path planning helps the vehicle to choose a safe, convenient, economically efficient path. Machine learning, part of AI, can learn from past experience, hence predicting the best path plan for the vehicle which can be accurate and safe.

The next step is to follow the path safely in order to execute the set plan of the path. For a safe drive, autonomous vehicles have to detect traffic objects such as cars, bicycles, pedestrians, traffic signs, and signals. The main inputs for autonomous vehicles are digital images, and vehicles must detect objects in digital images. With the help of deep learning algorithms, vehicles can predict the class and exact location of an object in a given image.

Vehicle monitoring: The vehicle monitoring system is an important component in autonomous vehicles. It can inform about the current condition of the vehicle and can predict the problems that may arise for the vehicle. For such type of prediction, machine learning algorithms can be used.

1.6 Benefits of Autonomous Vehicle

Improving traffic safety: There are many causes of vehicle crashes, like high speed, adverse weather, rash driving, drinking and driving, careless driving, taking naps during long drives. Vehicle crashes are the main cause of human casualties. In comparison to the human-driven vehicle, an autonomous vehicle could provide increased safety while traveling. Autonomous vehicles can reduce vehicle crashes because software installed in the vehicle instructs the autonomous vehicle's control system rather than humans, and software makes fewer errors than humans.

Easy transportation for physically handicapped and elderly people: An autonomous vehicle can be a facility for those who are not able to drive. They can use an autonomous vehicle for transportation without being a burden on others.

Follow traffic rules: Every year, road accidents occur as a result of violations of traffic rules, resulting in the loss of life as well as economic losses. Breaking traffic rules is a natural human tendency that is sometimes caused by circumstances and other times by a lack of knowledge of traffic rules. However, this can result in traffic accidents. Such vehicle accidents can be reduced by an intelligent traffic sign and signal recognition system in an autonomous vehicle.

Reduced traffic congestion: Due to human driving tendency, frequent lane change, stop and go, and rash driving, are the main cause of traffic congestion. Rash driving is a main cause of accidents which leads to traffic congestion. In traffic congestion, a vehicle generates more CO2 than normal traffic, and that impacts the environment as well, and there is also a loss of time and money. A lower number of crashes could also reduce traffic congestion. An autonomous vehicle can reduce the frequency of crashes because of inbuilt intelligent software that removes human driving behaviors like rash driving, and stop and go driving. Therefore, with an intelligent system, an autonomous vehicle can reduce traffic congestion as it follows traffic rules resulting in fewer accidents.

1.7 Applications of Autonomous Vehicle

Efficient parking: Car parking is one of the most challenging tasks. Almost every organization dedicates one-third of its land for parking. Autonomous vehicles can reduce parking problems because they can drop passengers in one location and park themselves away from the designated location.

Reduce Transportation Costs: Transportation costs also include driver fees along with fuel and other charges. Autonomous vehicles can reduce transportation costs because it does not require a driver.

Emergency warning system: The autonomous vehicle can detect accidents on roads and can notify emergency services and other vehicles.

Automatic speed control: The autonomous vehicle can automatically control the vehicle's speed according to the speed limit signal on the road. This feature can reduce the likelihood of vehicle accidents.

Transportation accessibility: Autonomous vehicles can reduce driving effort and can be easy and accessible transportation for seniors and people with disabilities.

Safe in night and in bad weather: An autonomous vehicle can be safe to drive at night and in bad weather. This is in contrast to human-driven cars, where drivers may become fatigued or exhausted, and may even go to sleep while driving, which may cause nighttime car accidents.

1.8 Anomalous Activities and Their Categorization

In autonomous vehicles, anomalous activity detection from video is a very challenging task. Safe driving requires anomalous activities to be detected in real time so that autonomous vehicles can make their own decisions at the right time and prevent road accidents. In the following section we describe possible types of anomalous activity in autonomous vehicles.

Hacking anomalies: In intelligent transport system, autonomous vehicles communicate with each other through internet. For a smooth and safe drive vehicles need to be connected to the internet, which is called Internet of Vehicles (IoVs). Once an autonomous vehicle is connected to the internet, it can be vulnerable to cyber-attacks. Cyber-attacks are a major anomalous activity that can be dangerous for safe driving in autonomous vehicles. Sniffing, denial of service and distributed denial of service are some examples of cyber-attacks. In IoVs, deep learning-based intrusion detection systems (IDS) can be helpful in detecting suspicious anomalous activity between vehicles.

Faulty sensor: Autonomous vehicles depend entirely on various inputs, such as LiDAR, RADAR, sensors, cameras, for information about the surroundings. If there is a fault with the sensor, it can cause abnormal behavior of autonomous vehicles and cause an accident.

Another vehicle behavior: The safety of a vehicle on the road also depends on how other vehicles in the vicinity are driving. There is always a possibility that the driver of a nearby vehicle may be intoxicated or not in good health. Driving near such vehicles can lead to road accidents. Such anomalous behavior of other vehicles should be detected in real time so that the self-driving car can maintain a safe distance from such vehicle.

False object detection: There is always a probability of false object detection in the image sequence. Advertising companies uses public transport to promote their products. For this, they paste product pictures on buses and taxis. Product images may also contain images of a person, animal, vehicle, etc. Another example is wet roads, which can also cause the image of one object to be visible on another vehicle. The deep learning model in autonomous vehicles should be trained in such a way that it should differentiate between actual objects and object images.

Road marking anomalies: Some traffic signs, such as zebra crossing, stop lines, and lane change marking (Rumble strips), are printed on roads. These road markings should be detected by vehicles in real time so that vehicles can take decisions about what action to take next. It may be that the road markings are covered with snow or water due to weather conditions. Such conditions or road marking anomalies not only impact autonomous vehicle driving but also affect human driving. Therefore, to detect such types of road marking anomalies is difficult in autonomous vehicles.

1.9 Deep Learning Methods in Autonomous Vehicle

Anomalies like roadside path holes, cracks, and other vehicle behavior may impact the normal safe drive of an autonomous vehicle. It can be useful to analyze the video stream to detect these types of anomalies. Videos are nothing but a continuous flow of still images. Recognition or detection of an object in a digital image is a computer vision problem. In recent years, authors proved that deep learning can be helpful in object classification, localization, and object detection. Convolutional neural network is the backbone of most object detection models. CNN family models such as R-CNN, Fast R-CNN, Faster R-CNN, and YOLO and SSD are the most popular object detection models.

R-CNN was proposed by R. Girshick *et al.* [19] in 2014. The selective search was used to propose a candidate region. Two thousand region proposals were extracted from the test images. In comparison to CNN, R-CNN uses a limited number of region proposals. AlexNet Deep CNN was used as a feature extractor. Using CNN, fixed-length feature was extracted from each region. Finally, linear SVM was used for the classification of each region. A regression model was used to predict the bounding box and to reduce the localization error. On multicore CPU R-CNN achieved 53.7 mAP in 10 seconds.

Fast R-CNN: R. Girshick *et al.* proposed an extension of their previous work R-CNN named Fast R-CNN. Fast R-CNN was faster than R-CNN. In Fast R-CNN, instead of 2,000 region proposals, the entire image along with the region of interest (ROI) was fed into the network. Pretrained CNN was used for feature extraction. The end of the network was the custom layer that was the ROI pooling layer to extract features from each region. Finally, each feature vector was processed by a fully connected layer. The output of the fully connected layer was divided into branches. Softmax layer was used for classification in one branch and another branch predicts coordinates for the object. The model achieved 66% mAP [20].

Faster R-CNN: Most CNN-based object detection algorithms used selective search to propose region proposals, Faster R-CNN Ren *et al.* on the other hand used a separate convolutional network, a region proposal network (RPN) for the region proposals. Both ConvNet and RPN share computation in the entire object detection process. For object classification in each region, the ROI pooling layer is used and the bounding box is predicted.

The role of deep learning in autonomous vehicles is not only about object detection, but also how agile the deep learning model is for accurate object detection. CNN family methods like R-CNN, Fast R-CNN, and Faster R-CNN are based on region proposals for object detection. Most of them use linear search to find the possible regions first then perform object detection. This can slow down the object detection process. Real-time applications like autonomous vehicles require object detection within milliseconds with accuracy so that the vehicle can take necessary action accordingly. Due to the agility in object detection of YOLO and SSD, deep learning models can be used in autonomous vehicles [21]. Faster R-CNN, SSD, and YOLOv2 were evaluated using mAP and other model evaluation methods like recall, precision; it was found that YOLO v2 and SSD are four times faster than Faster R-CNN in object detection. Between YOLOv2 and SSD, it was found that YOLOv2 achieved better mAP in comparison to SSD [22].

Originally YOLO was proposed by J. Redmon *et al.* [23] in 2016. It was a single neural network that can predict bounding boxes and class probabilities from an image. Proposed YOLO architecture is a unified architecture that is extremely fast and can detect objects at 45 frames per second, which can be considered detection in real time. YOLO architecture is shown in Figure 1.3, The YOLO framework has 24 convolutional layers and 2 fully connected layers.

The proposed YOLO has achieved 51-57% mAP on COCO benchmark dataset. But J. Redmon *et al.* have found some problems in original YOLO, such as, when compared to Faster R-CNN, there are localization errors in YOLO. YOLO achieved low recall as compared to the region proposal methods [23]. In 2017, J. Redmon introduced another version of YOLO, known as YOLO 9000 or YOLO v2. It was trained and tested in over 900 object categories and was capable of detecting over 900 objects in real time. Experimental results showed that YOLOv2 achieved 76.8 mAP at 67 FPS and 78.6 mAP at 40 FPS on VOC2007 [24]. In 2018, J. Redmon *et al.* proposed the third version of YOLO and named it YOLOv3. At 320 X 320 YOLOv3 achieved 28.2mAP. While considering at .5 IOU detection matrices YOLOv3 achieved 57.9 AP in 51 ms and 57.5 AP in 198 ms which was 3.8 times faster [25]. In 2020, YOLOv4 was proposed by Alexey Bochkovsky *et al.* [26]. This time J. Redmon was not part of the team. In experimental results, YOLOv4 performs the fastest and most accurate real-time object detection. YOLOv4 uses oF (bag of freebies) and several BoyS (bag of specials) for object detection. YOLOv4 achieved 43.5 AP on COCO dataset at 65 FPS. In 2020, after two months of YOLOv4 release, Glenn Jocher [27] released YOLOv5. YOLOv5 was implemented on PyTorch. Data augmentation and auto-learning bounding boxes were major improvements in YOLOv5. In the study [28, 30] it was found that deep learning methods can be used in autonomous vehicles for object detection in real time. YOLO and SSD can also be helpful in real-time traffic sign detection and recognition [29, 31].

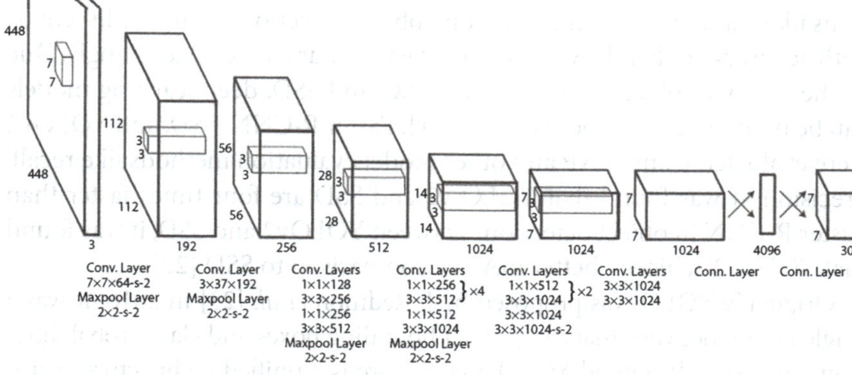

Figure 1.3 Original YOLO framework. (Source: [19].)

1.10 Working of Yolo

The entire image is cast into the network for classification and prediction of bounding box. The image is divided into SXS grid. A feature map is created from entire image. The feature map is used to predict a bounding box for each and every class. For each grid cell N a bounding box was generated with confidence score. Confidence score determines the presence and absence of object in a grid. Zero confidence score represents the absence of an object and if the class probability and bounding box is higher than a certain confidence score, ensures the presence of object (Figure 1.4 and Figure 1.5).

For each grid cell YOLO predicts output y. If there are four classes then the output y has 3X3X1X9 dimension where 3X3 is grid cell, 1 is number of bounding box and 9 is dimension. Pc represents whether object is present in the grid cell or not. bx,by bh,bw are bounding box coordinates and c1,c2,c3,c4 are classes. The PC value 0 indicates the absence of an object in that grid cell. If the Pc value is 0 then there is no need to calculate the bounding box and other values. Pc value 1, represents the presence of an object in the grid cell. Figure 1.5 depicts how YOLO predicts bounding box coordinates.

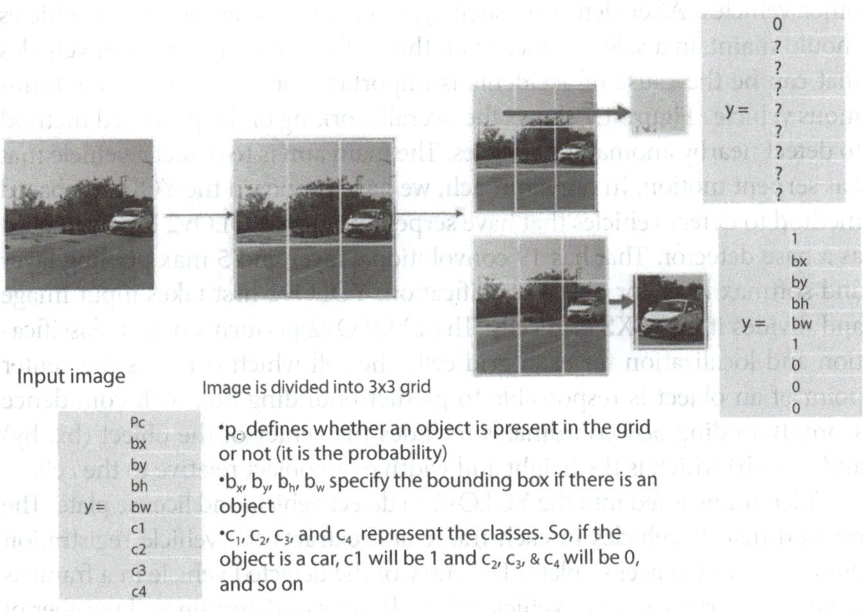

Figure 1.4 Prediction of bounding box for each grid cell.

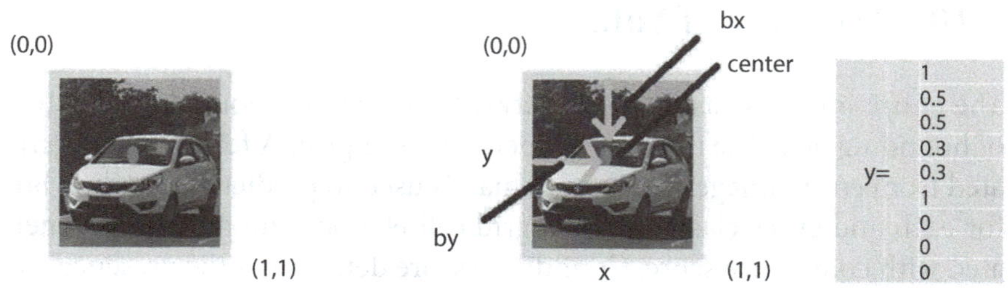

Figure 1.5 Coordinates bx,by,bh and bw specifies the bounding box.

1.11 Proposed Methodology

The serpent move of the nearby vehicle can be a danger for a safe drive in an autonomous vehicle. It can be considered as an anomalous activity that can affect the smooth drive-in autonomous vehicles. In this section, we are proposing a method to detect nearby vehicles which are driving in a serpent manner. Figure 1.6 shows serpent/turbulence drive of nearby vehicle.

In Figure 1.6 vehicle A is an autonomous vehicle, and vehicle B is a normal vehicle. It is clearly shown that vehicle B has a serpent motion. The camera at the front of the autonomous vehicle will capture all other vehicles that will come in front of the vehicle. Because of serpent motion, vehicle B will be considered as an anomalous vehicle that can be the cause of an accident for other vehicles. After detecting such types of vehicles, autonomous vehicles should maintain a safe distance from them. To detect such types of vehicles that can be the cause of accidents is important for a safe drive-in autonomous vehicle. Figure 1.7 shows the overall working of the proposed method to detect nearby anomalous vehicles. The main aim is to detect a vehicle that has serpent motion. In our approach, we have proposed the YOLOv2-based method to detect vehicles that have serpent motion. YOLOv2 has been used as a base detector. That has 19 convolutional layer and 5 max pooling layer and softmax layer for object classification. YOLOv2 first takes input image and divides it into SXS grid cells. Then YOLOv2 performs object classification and localization for each grid cell. The cell which contains the center point of an object is responsible to predict bounding box with confidence score. Bounding box coordinates includes the center of the object (bx, by) and (bw,bh) which is the height and width of an object relative to the cell.

Video frame is fed into the YOLOv2 to detect vehicle and license plate. The method detects vehicles in each frame and extracts the vehicle registration number from the license plate. The entry of the detected vehicle in a frame is done in the primary table. Vehicle ID (Vid), time of detection and number of

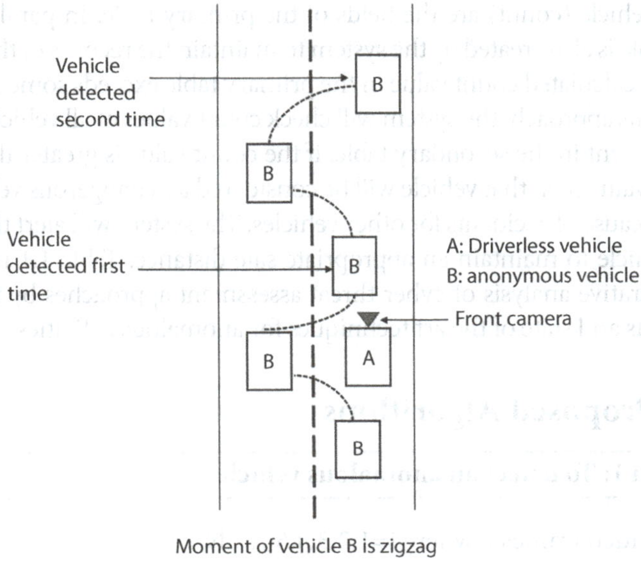

Figure 1.6 The zigzag drive of a nearby vehicle can be an unsafe drive for an autonomous vehicle.

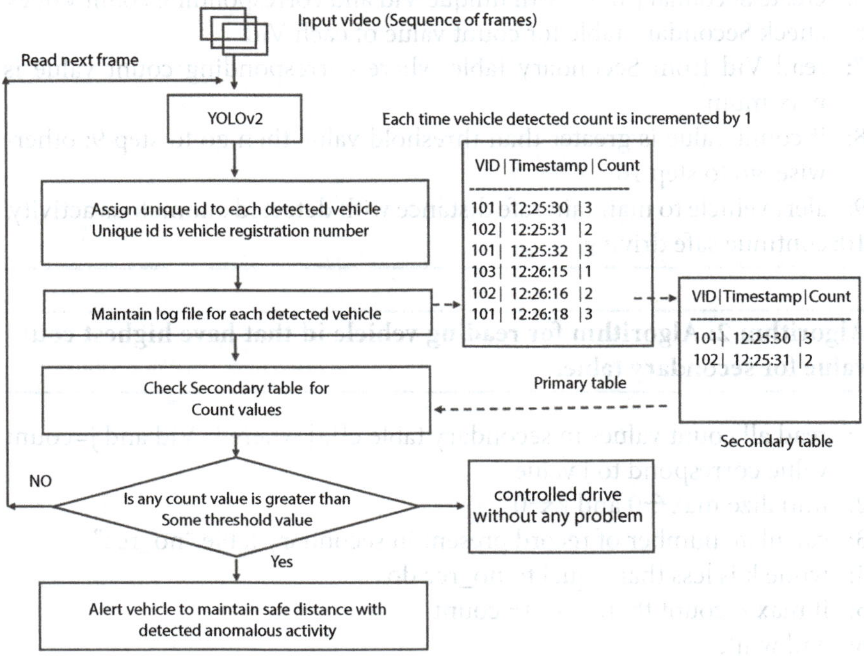

Figure 1.7 Flow of proposed method.

detected vehicle (count) are the fields of the primary table. In parallel, a secondary table is also created by the system to maintain the records of those vehicles whose calculated count value in the primary table exceeds some threshold value. In this approach, the system will check count values of all vehicles whose entry is present in the secondary table. if the count value is greater than some threshold value then that vehicle will be considered as a dangerous vehicle that can be the cause of accidents for other vehicles. The system will alert the autonomous vehicle to maintain an appropriate safe distance. Table 1.1 represents the comparative analysis of cyber threat assessment approaches by focussing on methods and state of the art techniques for anomalous activities.

1.12 Proposed Algorithms

Algorithm 1: To detect an anomalous vehicle

1: input video frames f_i where i=1,2,3,4,5 .... N
2: apply YOLOv2 model for vehicle detection
3: extract vehicle registration number "Vid" of detected vehicle
4: insert extracted vehicle number Vid in "Primary table" and increment corresponding count value, count++
5: create Secondary table with unique Vid and corresponding count values
6: check Secondary table for count value of each Vid
7: read Vid from Secondary table where corresponding count value is maximum
8: if count value is greater than threshold value then go to step 9; otherwise, go to step 10
9: alert vehicle to maintain safe distance with detected anomalous activity.
10: continue safe driving

Algorithm 2: Algorithm for reading vehicle id that have highest count value for secondary table.

1: read all count values in secondary table c[i,j] where i=Vid and j=count value correspond to i value
2: initialize max←0 and k←0
3: calculate number of record present in secondary table "no_rec"
4: while k is less than equal to no_rec do
5: if max < count then max ← count
6: end while
7: read Vid where max equal to count

1.13 Comparative Study and Discussion

Table 1.1 Comparative analysis.

Methods	Anomalous activity	Result
Security model based, a deep learning based network Deep denoising auto encoder [1]	Deal with cyber-attacks for ECU	Outperforms other ML models, achieved hit highest rate and lowest miss rates
mIoUNe Network [2]	Detecting failure cases in semantic segmentation	Achieved 93.21% mIoU accuracy & 84.8% failure detection
Automatic Pavement Distress Recognition (APDR) system [3]	Road anomalies like path hole cracks on roads	Able to perform road anomalies detection
Anomaly detection method using CNN and Kalman filtering [4]	Detection of anomalous reading in sensors	Achieved high accuracy, sensitivity, and F1 score
Federated sensor data fusion architecture [5]	Deal with sensor faults	Architecture can detect soft and hard faults from a particular sensor
Deep learning-based method [6]	To detect anomalous vehicles and reidentification of vehicles	Model can detect anomalous vehicles
Proposed highway traffic anomaly dataset [7]	Anomalous traffic incidents and patterns	CGAN, FlowNet, PredNet N+1, and PredNEt N +6 models were evaluated using the HTA dataset, PredNet N+ 6 achieved the highest AUC score among other models in detecting anomalies

(Continued)

Table 1.1 Comparative analysis. (*Continued*)

Methods	Anomalous activity	Result
Deep learning-based fault detection method [8]	Fault detection	Deep learning model classified signals into two categories faulty or not faulty
Decision tree-based method [9]	Detection of traffic anomalies	Achieved a 0.8571 F1 scores and 0.5686 S4 score
Unsupervised learning model based on deep autoencoder [10]	Detect the self-reported location anomaly	Detecting CAV location anomaly in a complicated scenario
Hidden Markov Model (HMM) based behavior analysis method to assess vehicles' driving behavior [15]	Detect anomalous pattern	Successfully detect the risk moments
Pixel-wise anomaly detection Framework [16]	Anomaly segmentation	Can be used with any pre-trained cutting-edge segmentation model
Deep learning method [17]	To avoid anomalies like signal interference, software and hardware errors, cyber-attacks	Accuracy obtained was 95.5% and precision of 94.2%
Two-stage method	Deal with traffic anomalies sudden brake of another vehicle.	Detect and recognize vehicle tail brake lights from the noisy images
Proposed method	Detect *Traffic-Induced Turbulence*/ abnormal behavior/ turbulence of traffic in highways	To detect vehicle and recognize an anomalous vehicle after reading highest count value from secondary table

1.14 Conclusion

Autonomous driving is self-driving with sensor-based assessment of traffic and surrounding conditions. While driving there are various anomalous activities that can interrupt and become an obstacle for self-driving. The technology behind these sensors-based autonomous driving relies on the use of deep learning models. We reviewed that there are various conditions, such as infrastructure, traffic, weather and software security that could be a cause of anomalies in autonomous vehicles. Real-time anomalous activity could exist in numerous types, such as cyber-attacks, road anomaly (path holes, cracks, humps and bumps), sensors fault, false detection and other vehicle inappropriate movement, which may become the reason for road accidents. Serpent motion or turbulence of other vehicles can also be the cause of road accidents. This anomalous activity of the nearby vehicle can be a reason for the accident of the autonomous vehicle. Automatic detection of such a type of vehicle is also important. To avoid these anomalies and for safe driving, a number of deep learning models have been applied till now. In our study, we have proposed YOLOv2-based method that can detect such types of anomalous vehicles. We have trained YOLOv2 on the custom dataset and have found that YOLOv2 is state-of-the-art and faster than other RCNN-based detection models. We observed that for real-time anomalies detection, the latest version of YOLOv2 may be a most promising model.

References

1. Lin, Y., Chen, C., Xiao, F., Avatefipour, O., Alsubhi, K., & Yunianta, A. (2020). An Evolutionary Deep Learning Anomaly Detection Framework for In-Vehicle Networks-CAN Bus. *IEEE Transactions on Industry Applications*.
2. Song, J., Ahn, W., Park, S., & Lim, M. (2021). Failure Detection for Semantic Segmentation on Road Scenes Using Deep Learning. *Applied Sciences*, 11(4), 1870.
3. Tedeschi, A., & Benedetto, F. (2017). A real-time automatic pavement crack and pothole recognition system for mobile Android-based devices. *Advanced Engineering Informatics*, 32, 11–25.
4. Van Wyk, F., Wang, Y., Khojandi, A., & Masoud, N. (2019). Real-time sensor anomaly detection and identification in automated vehicles. *IEEE Transactions on Intelligent Transportation Systems*, 21(3), 1264–1276.

5. Realpe, M., Vintimilla, B., & Vlacic, L. (2015). Sensor fault detection and diagnosis for autonomous vehicles. In *MATEC Web of Conferences* (Vol. 30, p. 04003). EDP Sciences.
6. Khorramshahi, P., Peri, N., Kumar, A., Shah, A., & Chellappa, R. (2019, June). Attention Driven Vehicle Re-identification and Unsupervised Anomaly Detection for Traffic Understanding. In *CVPR Workshops* (pp. 239–246).
7. Singh, H., Hand, E. M., & Alexis, K. (2020, October). Anomalous Motion Detection On Highway Using Deep Learning. In *2020 IEEE International Conference on Image Processing (ICIP)* (pp. 1901–1905). IEEE.022 0.
8. Ren, J., Ren, R., Green, M., & Huang, X. (2019, August). A deep learning method for fault detection of autonomous vehicles. In *2019 14th International Conference on Computer Science & Education (ICCSE)* (pp. 749–754). IEEE.
9. Aboah, A. (2021). A vision-based system for traffic anomaly detection using deep learning and decision trees. In *Proceedings of the IEEE/CVF Conference on Computer Vision and Pattern Recognition* (pp. 4207–4212).
10. Wang, X., Mavromatis, I., Tassi, A., Santos-Rodriguez, R., & Piechocki, R. J. (2019, September). Location Anomalies Detection for Connected and Autonomous Vehicles. In *2019 IEEE 2nd Connected and Automated Vehicles Symposium (CAVS)* (pp. 1–5). IEEE.
11. Alheeti, K. M. A., Gruebler, A., & McDonald-Maier, K. (2017). Using discriminant analysis to detect intrusions in external communication for self-driving vehicles. *Digital Communications and Networks,* 3(3), 180–187.
12. Ali Alheeti, K. M., & McDonald-Maier, K. (2018). Intelligent intrusion detection in external communication systems for autonomous vehicles. *Systems Science & Control Engineering,* 6(1), 48–56.
13. Berlin, O., Held, A., Matousek, M., & Kargl, F. (2016, December). Poster: Anomaly-based misbehaviour detection in connected car backends. In *2016 IEEE Vehicular Networking Conference (VNC)* (pp. 1–2). IEEE.
14. Rajbahadur, G. K., Malton, A. J., Walenstein, A., & Hassan, A. E. (2018, June). A survey of anomaly detection for connected vehicle cybersecurity and safety. In *2018 IEEE Intelligent Vehicles Symposium* (IV) (pp. 421–426). IEEE.
15. Yang, C., Renzaglia, A., Paigwar, A., Laugier, C., & Wang, D. (2019, November). Driving behavior assessment and anomaly detection for intelligent vehicles. In *2019 IEEE International Conference on Cybernetics and Intelligent Systems (CIS) and IEEE Conference on Robotics, Automation and Mechatronics (RAM)* (pp. 524–529). IEEE.
16. Di Biase, G., Blum, H., Siegwart, R., & Cadena, C. (2021). Pixel-wise Anomaly Detection in Complex Driving Scenes. In *Proceedings of the IEEE/CVF Conference on Computer Vision and Pattern Recognition* (pp. 16918–16927).
17. Oucheikh, R., Fri, M., Fedouaki, F., & Hain, M. (2020). Deep real-time anomaly detection for connected autonomous vehicles. *Procedia Computer Science*, 177, 456–461.
18. Wang, J. G., Zhou, L., Pan, Y., Lee, S., Song, Z., Han, B. S., & Saputra, V. B. (2016, June). Appearance-based brake-lights recognition using deep

learning and vehicle detection. In *2016 IEEE Intelligent Vehicles Symposium* (IV) (pp. 815–820). IEEE.
19. Girshick, Ross, Jeff Donahue, Trevor Darrell, and Jitendra Malik. "Rich feature hierarchies for accurate object detection and semantic segmentation." In *Proceedings of the IEEE Conference on Computer Vision and Pattern Recognition*, pp. 580–587. 2014.
20. Girshick, Ross. "Fast r-cnn." In *Proceedings of the IEEE International Conference on Computer Vision*, pp. 1440–1448. 2015.
21. Ren, Shaoqing, Kaiming He, Ross Girshick, and Jian Sun. "Faster r-cnn: Towards real-time object detection with region proposal networks." *Advances in neural information processing systems* 28 (2015): 91–99.
22. Garg, Priya, Debapriyo Roy Chowdhury, and Vidya N. More. "Traffic sign recognition and classification using YOLOv2, Faster RCNN and SSD." In *2019 10th International Conference on Computing, Communication and Networking Technologies (ICCCNT)*, pp. 1–5. IEEE, 2019.
23. Redmon, J., Divvala, S., Girshick, R., & Farhadi, A. (2016). You only look once: Unified, real-time object detection. In *Proceedings of the IEEE conference on computer vision and pattern recognition* (pp. 779–788).
24. Redmon, J., & Farhadi, A. (2017). YOLO9000: better, faster, stronger. In *Proceedings of the IEEE conference on computer vision and pattern recognition* (pp. 7263–7271).
25. Redmon, J., & Farhadi, A. (2018). Yolov3: An incremental improvement. arXiv preprint arXiv:1804.02767.
26. Bochkovskiy, A., Wang, C. Y., & Liao, H. Y. M. (2020). Yolov4: Optimal speed and accuracy of object detection. arXiv preprint arXiv:2004.10934.
27. Glennjocher *et al.* yolov5. https://github.com/ ultralytics/yolov5, 2021.
28. Amit Juyal, Sachin Sharma, and Priya Matta. "Deep Learning Methods for Object Detection in Autonomous Vehicles." In *2021 5th International Conference on Trends in Electronics and Informatics (ICOEI)*, pp. 751–755. IEEE, 2021.
29. Amit Juyal, Sachin Sharma, and Priya Matta. "Traffic Sign Detection using Deep Learning Techniques in Autonomous Vehicles." In *2021 International Conference on Innovative Computing, Intelligent Communication and Smart Electrical Systems (ICSES)*, pp. 1–7. IEEE, 2021.
30. Piyush Agarwal, Sachin Sharma, and Priya Matta. "Components, Technologies, and Market of Road Traffic Management System in Global Scenarios: A Complete Study." In *2021 International Conference on Innovative Computing, Intelligent Communication and Smart Electrical Systems (ICSES)*, pp. 1–10. IEEE, 2021.
31. Anuj Singh, Sachin Sharma, Kamlesh Chandra Purohit, and KC Nithin Kumar. "Artificial Intelligence based Framework for Effective Performance of Traffic Light Control System." In *2021 International Conference on Innovative Computing, Intelligent Communication and Smart Electrical Systems (ICSES)*, pp. 1–4. IEEE, 2021.

2

Algorithms and Difficulties for Autonomous Cars Based on Artificial Intelligence

Sumit Dhariwal[1]*, Avani Sharma[1] and Avinash Raipuria[2]

[1]Department of Information Technology, Manipal University, Jaipur, India
[2]Department of Information Technology, Malaysia University of Science & Technology, Petaling Jaya, Selangor, Malaysia

Abstract
In the past, no one could have imagined that automobiles would eventually be able to drive themselves, but with the help of science and technology, autonomous vehicles are now a realistic alternative in a variety of industries. Agriculture, transportation, and the military are just a few examples. Vehicles must gather information, determine routes, and carry out their plans. To move, these vehicles require sensor data and artificial intelligence (AI) algorithms. Machine learning techniques are employed and this is part of AI. This two-part essay analyses the AI solicitation that will allow self-driving vehicles to become a reality, as well as the issues and solutions that will be encountered.

Keywords: Route planning, object detection, and decision making, support vector machine

2.1 Introduction

In some sections of industry, independent vehicles [1] are starting to become a serious prospect. Agriculture, transportation [2], and defense are only a few examples. The day when autonomous vehicles will be seen in everyday life by the average person is rapidly coming. Sensor data and artificial intelligence (AI) algorithms [3] are used in many of the functions that cars have. Vehicles must acquire data and plan their routes, and then, and only then, should you follow those directions. These tasks necessitate

**Corresponding author*: sumit.dhariwal@jaipur.manipal.edu

Romil Rawat, A. Mary Sowjanya, Syed Imran Patel, Varshali Jaiswal, Imran Khan and Allam Balaram (eds.) Autonomous Vehicles Volume 1: Using Machine Intelligence, (27–38) © 2023 Scrivener Publishing LLC

the use of non-traditional software development [4] approaches and AI techniques such as machine learning. This two-part series delves into the AI apps that help autistic people.

The first attempts to automate vehicle control date back to the 1920s. In 1926, the Chandler Company launched a radio-controlled vehicle guided by an operator in a vehicle that followed close behind. Of course, this was not about vehicle autonomy, but it was the first time a car had been driven without a crew on the road [5, 6].

In 1956, General Motors fitted an experimental Firebird II model with a receiver for directing signals supplied through a wire placed in the highway of the future [7]. Using an image processing method named The Karts vision algorithm, an experimental Stanford cart was able to travel in a closed space without human interaction in 1979. This approach was based on the block world planning method and involves reducing the image to a collection of edges; however, it was found to be inappropriate for outdoor usage due to the presence of numerous complex forms and colors [8].

The Blocks World technique from the 1960s, on the other hand, is currently one of the most well known in the domains of artificial intelligence planning [9] and environmental recognition [10, 11].

The VMP vehicle, which was the first autonomous automobile capable of traveling about a defined region, covered a thousand kilometers without human aid in 1995. The prototype could navigate through traffic and pass cars, and the EMS-Vision autonomous system was built using data from a biaxial political platform with a bifocal camera system (45 and 15 degrees) [12]. High-speed driving, the ability to spot possible dangers across long and short distances, and the ability to understand uneven terrain and spatial relationships were all factors in deciding which cameras to utilize. (More current versions of the EMS-Vision system use a system of three cameras [13]).

The following were utilized by EMS-Vision:

Statistical data; road network maps; permanent items on the map that were utilized as markers (waypoints), such as traffic lane width.

Developing dynamic knowledge, which includes gathering information about the surroundings while driving, recognizing objects (markers), and computing the position (location and orientation) of things in space using HTC (Homogeneous Coordinate Transformation) algorithms.

There were further decision-making units for keeping key things in the field of vision (to be examined by specific modules), driving vehicles, and a central decision-making unit in the system. In the event of a lack of work, disputes between the control and care units, or other issues, the central unit prioritized decision-making.

EMS-Vision includes the following modules:

- The algorithm not only identified the path but also generated a model from the connected segments and used it to calculate the vehicle's position.
- Controls attention and communicates with a two-axis rotary head that executes jump (saccadic) motions in relation to an item.

The results of the calculations were passed to the central unit, which chose the path. The chosen path was added to the mission plan's task list. The mission plan contained duties including determining the best path, tracking things (roads, landmarks), and more.

Vehicle control (steering): the module was responsible for analyzing information from decision-making units and performing algorithms for feedback control, feedforward control, and transitional rules; The module calculated the correcting variables and evaluated the resulting sets of equations.

Tesla introduced the S model in 2014 with Autopilot Firmware 7.0 software, which added a second degree of automation [16]. Table 2.1 shows the

Table 2.1 SAE level of automation [17].

Level	Name	Description
0	No Automation System	Complete lack of automation. All decisions and actions are taken by the driver.
1	Driver Assistance System	Driver assistance system for example in accelerate based on external information.
2	Partial Automation System	The system of level 1 is assumed autonomous, but the driver is obliged to react when he or she considers the decision of the automatic system to be incorrect.
3	Conditional Automation System	The vehicle is driven by automatic autonomous but the driver for the environment.
4	Conditional Automation System	The vehicle is driven by automatic systems, but the driver is obliged to react when the driver does not respond.
5	Full Automation System	This system is to perform for vehicle and is able for all driving tasks under any conditions and driver has no control of vehicle.

features of the SAE International (previously SAE - Society of Automotive Engineers) automation levels.

2.1.1 Algorithms for Machine Learning in Autonomous Driving

Autonomous vehicles are inextricably linked to the industrial Internet of Things (IoT) [14]. The IoT, in conjunction with other knowledge like machine learning, artificial intelligence, local computing, and others, is providing the technology required for autonomous vehicles. Many individuals are interested in learning how these self-driving automobiles function [15]. What is going on inside so they can operate without the need for drivers to take the wheel? It is common knowledge that modern automobiles have numerous devices, as well as actuators and controllers [16]. These are the final products controlled by software running on the ECU's different function-specific software (Electronic Control Unit) [17]. This bundle also includes machine learning software.

In self-driving cars, ML (machine learning) is vital because it constantly presents the surrounding environment and anticipates potential changes. The four subtasks of machine learning in self-driving cars are object detection, object identification or recognition, object categorization, and object localization and movement prediction. Algorithms for machine learning may be categorized into the following groups:

- The use of regression algorithms
- Recognizing patterns
- Algorithms for clustering and decision matrices.

The following are examples of regression techniques that might be utilized in self-driving cars:

- Bayesian regression is a technique for predicting the outcome of a situation
- Regression using neural networks
- Regression using decision trees.

2.1.2 Regression Algorithms

Images (radars or cameras) play a crucial role in A-D-A-S. In addition to localization and implementation, image-based models for prediction and feature selection must be developed. This is the toughest hurdle used for self-driving vehicles or any algorithm. Methods generally such as Bayesian regression, neural network regression, and decision forest regression are examples of self-driven regression approaches.

2.1.3 Design Identification Systems (Classification)

The data captured by the Semi-autonomous driving sensor contain a variety of environmental data; image filtering is required to detect instances of a certain item category by deleting redundant data points. Based classification techniques simply discard abnormal data points. It is crucial to discover trends in data collection for classifying things. This method is also known as a data minimization algorithm.

These techniques assist decrease data sets by identifying object edges then fitting line segments (changes are taking place) and circular arcs at the edges. After the line segments have been aligned, from edge to corner, a new line segment is initiated. For sequences, circular arcs are ideal.

Support Vector Machines (SVMs) with Scale-Invariant feature Gradient (HOG) and Feature Extraction are the most commonly utilized recognition methods in ADAS (PCA). In addition, the K Nearest Neighbour (KNN) and Validation Set Rule are used.

2.1.4 Grouping Concept

The organization's visuals are not constantly crisp, and things might be challenging to notice and find. It is also feasible that the categorization algorithm will examine the object, declining to categorize and report it to the system. This might be due to low-motion photos, a lack of data points, or data that is unbalanced. This system excels in the variables that are being used to extract the framework. In the same way, the regression does, it describes the class of the issue and the class of methods. Centroid-based and complex clustering algorithms are extensively used modeling methods. All of the approaches are centered on utilizing data frameworks to better arrange the data groups of maximal matching.

2.1.5 Decision Matrix Algorithms

This algorithm excels at methodically identifying, analyzing, and rating the performance of relationships between sets of values and data. These algorithms are mostly used to make decisions. Depending on the algorithm's confidence level in the categorization, identification, and prediction of the upcoming movement of objects, the automobile must turn left or apply the brakes. These algorithms are made up of many decision models that have been trained individually and whose predictions are integrated in some way to generate a single forecast while lowering the chance of making a mistake. Gradient boosting (GDM) and Ada boosting are the most widely utilized algorithms.

2.2 In Autonomous Cars, AI Algorithms are Applied

2.2.1 Algorithms for Route Planning and Control

For this objective, conventional computer science methods that are heuristic in nature can be utilised. These include algorithms such as Bellman-Ford and Dijkstra's [4]. To make these algorithms function, we need to keep the car localized at all times. Simultaneous localization and mapping (SLAM) methods, as well as sensors like GPS, are used to achieve localization.

When GPS is unavailable, such as underground or in confined areas, SLAM is employed. SLAM maps the environment while also estimating the vehicle's location [5]. To portray the surroundings, the map is made up of landmarks or barriers. In cases where a map is not available, SLAM is employed.

2.2.2 Method for Detecting Items

In a moving vehicle, object detection is one of the most critical tasks that AI must accomplish. These algorithms are a hot topic of study, and they rely on a variety of sensors. Cameras, lidars, radars, and other forms of sensors can be used to detect objects. The algorithms are usually deep learning algorithms that use a neural network to complete the task [13].

One of the requirements for this assignment is that it be completed as promptly as possible. The reason for this is that when the vehicle is moving, a series of photos must be analyzed.

Here are some of the most cutting-edge techniques based on Convolutional Neural Networks (CNN). R-CNN, Fast R-CNN, and You Only Look Once (YOLO) are the three approaches [13]. R-CNN initially identifies regions in an image that may contain probable objects, then attempts to evaluate each region. This slows down R-CNN to some extent, which is why quicker R-CNN approaches have been created, as well as the

YOLO method. Using the same convolutional neural network, YOLO collaborates to locate regions and classify items into regions.

2.2.3 Algorithmic Decision-Making

The vehicle's actions are determined by decision-making, which is based on the information received from the sensors. A vehicle makes decisions all of the time based on its policies and surroundings. The following are the algorithms that are utilized to make decisions [4]:

- Decision Trees
- Support Vector Machine (SVM) Regression
- Deep Reinforcement Learning

2.3 AI's Challenges with Self-Driving Vehicles

Some of the challenges of using artificial intelligence algorithms for autonomous vehicles are the same challenges that are universal for many other AI applications. The autonomous vehicles domain introduces some additional, unique challenges. These are concepts like real time, safety, and machine ethics. The following subsections describe these areas in detail.

2.3.1 Feedback in Real Time

Real-time systems are a subset of embedded systems that have the ability to generate output or feedback within a set amount of time. They must be designed to be deterministic and minimal in order to always meet the desired real-time behavior. They bypass interpreted languages and dynamic memory allocation by using customized real-time operating systems (RTOS) or bare metal authorities that communicate directly with the hardware. Real-time systems use a subset of the programming language to guarantee speed and determinism in some circumstances.

Artificial intelligence (AI) solutions are typically on the other end of the spectrum, relying on technology.

Another consideration is whether these systems are centralized or dispersed [6]. Architecting centralized systems are easier, but they rely significantly on internal communications and a powerful central processor. Dedicated CPUs that can handle several subsystems and sensors are used in distributed systems, minimizing the requirement for a complicated and powerful central CPU. Systems with reduced power consumption are possible because of the distributed architecture.

They are also more adaptable and may be less expensive.

2.3.2 Complexity of Computation

Artificial intelligence algorithms, especially deep learning, require specialized hardware solutions due to the amount of data and complexity of calculations.

Hardware, such as Graphics Processing Units (GPU) or Tensor Processing Units (TPU) can be highly optimized for fast parallel computation. Speed comes at the expense of higher energy consumption and higher cost. Even after using specialized hardware, there is still no certainty that particular algorithms will be able to reach solutions in real-time constraints. Therefore the selection of algorithms in relation to their complexity and CPU demands is a serious factor in real-time systems. CPU computation and memory usage are measures that can help determine whether an algorithm is suitable for use in the specific CPUs used in modern vehicles.

2.3.3 Black Box Behavior

AI algorithms have been chastised for being more difficult to decipher than traditional computer algorithms. This is due to the fact that AI algorithms are extremely sophisticated and rely on a large amount of data. Without knowing the process it governs, a sophisticated neural network can execute a variety of AI tasks.

In reality, a deep neural network approximates a function, which is inherent in its design and explains why it is so effective. Thousands of nodes can be used to train deep neural networks to reach a specific state. They can also have several hidden layers as well as multiple inputs and outputs. All of this causes complications when a system crashes, creating the necessity for forensic examination.

One option is to employ hybrid solutions that integrate artificial intelligence and classical control techniques [1]. This is a critical issue for accident prevention and management [10]. An AI algorithm must justify the decision history of one decision versus another, which is difficult or impossible to do if the algorithm can only be evaluated and regarded as a black box.

2.3.4 Precision and Dependability

Autonomous car computer vision applications may not be ready for prime time. The reason for this is that they can work under perfect conditions but can also fail if the sensor inputs are messed up. AI algorithms are gradually trained using training data with certain properties.

Changing the data might cause categorization and prediction algorithms to behave in unexpected ways, with potentially disastrous results. A person with a large bag, for example, cannot be treated as a person by a vehicle. Furthermore, it is difficult to forecast what will happen when random data enters the system. Assailants have taken advantage of the flaw.

2.3.5 The Safeguarding

AI's complexity can lead to security concerns. It is harder to build, test, and deploy a more complicated system. This is merely one facet of the issue. Another truth is that new safety standards are emerging, but the transportation industry has yet to accept them [6].

What makes matters even more challenging is that AI system analysis and validation are still in their early phases and are difficult to track due to their complexity. Significantly for typical software systems, formal verification is tough, and it is even more difficult for AI-based software. The difficulty is in developing a reliable, verifiable system.

2.3.6 AI and Security

Because AI systems are so important to autonomous vehicles, their safety is directly tied to the system's overall reliability. Machine learning systems are vulnerable to attacks, as evidenced by several incidents [11].

The greatest risk arises when an opponent is able to falsify data flowing into the car through sensors, causing the system to make incorrect judgments. Training systems with data with minimal adverse qualities can improve this behaviour. "Just putting a few black and white stickers to a stop sign made the computer recognize it was a 45mph speed limit," according to a deep learning model for detecting traffic signs.

Sensor operation, which causes changes in the sensor data stream and thus entirely confuses AI systems, is a big danger to how AI performs in an autonomous vehicle.

Researchers are experimenting with solutions such as model-based hybrid systems, fuzzy logic, and classical control systems to augment AI.

The stakes are significant since artificial intelligence (AI) is now present in some vehicle subsystems and will soon be in every vehicle.

2.3.7 AI and Ethics

The ethical implications of AI deployment in autonomous and semi-autonomous cars are now undeveloped and undeveloped. Moral worth is a human characteristic that is difficult to formalize and apply to technology. To put it another way, a machine's ability to determine what is right or wrong is somewhat vague [10]. It may take decades for machine ethics to fully mature. It's an active study area that straddles the fields of human psychology, machine learning, and social policy. As the technology matures, it's feasible that some legislation will assist in this direction.

2.4 Conclusion

Today, AI solutions can be found everywhere. They are in devices like Alexa and Google Home, as well as the grass-cutting robots and driverless vehicles we'll be rolling out soon. AI has difficulties and successes in many industries, but the transportation industry, especially with autonomous vehicles, faces additional challenges related to safety and security. This is due both to the difficulty of developing a self-driving vehicle and to the state of AI research and its ability to address complex problems such as object recognition, route planning and decision making in real time. As we progress, many of these obstacles will be removed.

References

1. Abduljabbar, R., & Dia, H. (2019). Applications of Artificial Intelligence in Transport: An Overview. *Sustainability, 1*(11), 189.
2. Beachler, R. (2019). How Man-Made Neural Networks Emulate Biology to Advance AI and ML. *Electronic Design.*
3. Boucherat, X. (2019). Suppliers Forge the Relationship between AI and Sensors. *Automotive World.*
4. Bugala, M. (2018). Algorithms Applied in Autonomous Vehicle Systems. *Szybkobiene Pojazdy Gasienicowe, 50*(4), 119–138.
5. Cadena, C., Carlone, L., Carrillo, H., Latif, Y., Scaramuzza, D., Neira, J., Reid, I., & Leonard, J. (2016). Past, Present, and Future of Simultaneous

Localization and Mapping: Towards the Robust-Perception Age. *IEEE Transactions on Robotics, 32*(6), 309–1332.
6. Carmody, Thomas. (2019). Whitepaper: AI in the Driving Seat: Key Challenges for the Future of Advanced Driver Assistance. *Cambridge Consultants.*
7. Carol, T. (2017). Insurance and the Evolution of Automated Driving Systems. *National Association of Mutual Insurance Companies*, 2–11.
8. Khayyam, H., Javadi, B., Jalili, M., & Jazar, R. (2019). Artificial Intelligence and Internet of Things for Autonomous Vehicles. *Nonlinear Approaches in Engineering Applications (Book)*, 39–68.
9. Lentin, J. (2017). *Ros robotics projects*. Packt Publishing.
10. Narayanan, A. (2019). Ethical Judgement in Intelligent Control Systems for Autonomous Vehicles. *Australian New Zealand Control Conference (ANZCC), Auckland, New Zealand*, 231–236.
11. Newman, J. (2019). Whitepaper: Toward AI Security – Global Aspirations for a More Resilient Future. *CLTC White Paper Series.*
12. Prytz, R. (2014). Machine learning methods for vehicle predictive maintenance using off-board and on-board data. Halmstad University Dissertations no. 9, 1–32.
13. Redmon, J., Divvala, S., Girshick, R., & Farhadi, A. (2016). You Only Look Once: Unified, Real-Time Object Detection. *IEEE Conference on Computer Vision and Pattern Recognition (CVPR), Las Vegas, NV*, 779–788.
14. Sumit Dhariwal *et al.* (2020). "Image Normalization and Weighted Classification using and Efficient Approach for SVM Classifiers," International Journal of Image and Graphics (**World Scientific**). Vol.20, Issue no. 4, article No.2050035, 2020.
15. Sumit Dhariwal *et al.* (2019). "An Efficient Approach for Semantic Image Classification using Normalization Method" in *Advanced Trends in Computer Science and Engineering*, Vol. 8(4), July-August 2019, pp. 1268-1274 Volume 8, No.4, July – August 2019, *International Journal of Advanced Trends in Computer Science and Engineering.*
16. Sumit Dhariwal *et al.* (2018). "A New Way of Grace Assortment using Normalized Feature Value by Support Vector Machine " in *European Journal of Scientific Research,* ISSN 1450-216X/1450-202X Vol. 151 No 2 December 2018, pp. 197-204.
17. Sagar, V., & Nanjundeswaraswamy, T. (2019). Artificial Intelligence in Autonomous Vehicles – A Literature Review. *I-manager's Journal of Future Engineering & Technology*, *14* (3), 56–62.

3

Trusted Multipath Routing for Internet of Vehicles against DDoS Assault Using Brink Controller in Road Awareness (TMRBC-IOV)

Piyush Chouhan* **and Swapnil Jain**

Department of Electrical and Electronics Engineering, Shri Vaishnav Vidyapeeth Vishwavidyalaya, Indore, India

Abstract

Vehicular Ad-hoc network is described as communication of data or information between vehicles. This network can also support vehicle to infrastructure, pedestrian or other vehicle communication. Various applications of VANET are Safety Oriented Applications, Collision Evasion, Cooperative Driving, Traffic Improvement and Infotainment Applications. Advancement in technology created the new era known as Internet of Vehicles (IoV) which connects all of network vehicles to the internet, and it is gaining popularity because it offers significant benefits over previous works. IoT is based on the concept of connecting vehicles to communicate via the internet. These vehicles have internet-connected sensors that are individually identifiable and communicate with one another to execute complicated activities. The active type of assaults that lowers network performance is distributed denial of service; such assaults have resulted in significant losses by disrupting vehicular network functionality. Distributed Denial of Service (DDoS) assaults is one of the most dangerous threats to network functionality among all network assaults. Security challenges such as DDoS assaults have attracted the attention of stakeholders in the VANET environment to provide substantial solutions. With the increasing interconnectivity of vehicles and the lack of an entity preventing communications in IoV, security challenges such as DDoS assaults have become a critical challenge. Although similar issues exist in other contexts, their impact in the IoV environment might be more damaging, which is concerning since it allows intruders to enter the ecosystem in VANET. These assaults have a number of repercussions that might result in information loss in the vehicular network and other

**Corresponding author*: pchouhan06@yahoo.com

Romil Rawat, A. Mary Sowjanya, Syed Imran Patel, Varshali Jaiswal, Imran Khan and Allam Balaram (eds.) Autonomous Vehicles Volume 1: Using Machine Intelligence, (39–60) © 2023 Scrivener Publishing LLC

ad hoc wireless sensor networks. In this work, a key agreement approach is developed that identifies and isolates malignant vehicles from the network with the help of proposed trust management self-system called extension of trusted multipath distance vector routing (TMDR-Ext) which works on trusted authority based on brink controller (BC) and trusted threshold level, therefore improving the session recovery technique and reducing execution time. In terms of specific parameters, the suggested approach has been executed and compared to current techniques on the basis of QoS metrics, i.e., throughput, routing overhead, and packet delivery ratio are the three parameters. The suggested technique generates traffic with the help of Simulation of Urban Mobility tool (SUMO), and analysis and simulation of the proposed work is done by Network Simulator 2 (NS-2) which delivered good quality of service that indicates the new method outperforms current assessed results.

Keywords: VANET, IoV, DDoS attack, trust management system, SUMO, BC, TMDR-Ext, QoS and NS-2

3.1 Introduction

Wireless Ad-hoc organize is characterized as a system which does not have a prior correspondence framework [1]. The system is made by a few nodes which are accessible. In this type of network, determination of which nodes to transfer data to which node is done dynamically, depending upon the connectivity of both devices. Ad-hoc network can use flooding data transfer [2, 3]. In ad-hoc network can utilize flooding information exchange. In Ad-hoc arrange all gadgets are dealt with similarly and all have the same status. The fundamental utilization of remote specially appointed system is finished by Mobile Ad-hoc Network (MANET) [3]. In MANET different participating node moves randomly in the made wireless Ad-hoc system.

Mobile Ad-Hoc Network

We realize that in MANET nodes move arbitrarily in the wireless Ad-hoc network. Since here devices move progressively in the system, the interfacing join between two devices changes frequently, which makes building MANET very challenging. The main challenge of building MANET is to manage continuous change of route traffic.

Vehicular Ad-hoc network in late-year vehicular ad-hoc network (VANET) is a hot research subject. In VANET moving nodes are vehicle proceeding onward street. VANET can be utilized to enhance safety in the street [5].

Among the different kinds of MANET, the most famous is Vehicular ad-hoc network (VANET), which is utilized for communication between various vehicles running on traffic and street side framework [6]. We can likewise say that VANET is a subset of MANET where communication nodes are principally vehicles.

Vehicular Ad-Hoc Network

Vehicular ad-hoc network is made by applying the models of mobile ad-hoc network – the unconstrained creation of a wireless system for data exchange – to the zone of vehicles. VANETs were first exhibited in 2001 under "auto-to-auto extraordinarily named adaptable correspondence and system administration" applications, where the system can be surrounded and information can be given off among autos [7]. It was shown that vehicle-to-roadside and vehicle-to-vehicle exchanges structures will exist together in VANETs to give road prosperity, course, and other roadside administrations [7, 8]. VANETs are a key bit of the information transportation structures (ITS) framework. At times, VANETs are suggested as Intelligent Transportation Networks.

Nowadays, most current autos have an intra vehicular system which permits wireless communication amongst vehicle and electronic devices like smart mobile phone, Global Positioning System (GPS), and Bluetooth media players [4]. Be that as it may, the Inter vehicular ad-hoc network is as yet not accessible. So to give inter vehicular communication, VANET, i.e. Vehicular ad-hoc Network advances are rising. Vehicular ad-hoc (VANETs) are characterized a as a subset of mobile ad-hoc networks (MANETs) with the recognizing property that the nodes show in here are vehicles [6]. So, node, i.e., vehicle movement, is confined by street course, including movement and activity directions. Due to these limitations, VANET is supported by some fixed infrastructure that helps with a few services of the VANET and gives access to stationary systems. The fixed frameworks are sent to basic areas like roadsides, service stations, unsafe crossing points or places with dangerous climate conditions.

VANETs are rapidly growing in popularity due to their superior solutions for important and remote applications with the help of IoV. However, getting important remote data from vehicles is exceedingly challenging owing to limited range, memory, processing, and power supply [1, 2]. The use of ad hoc networks and radio waves for data transmission has also increased the likelihood of such networks being attacked [3, 4]. Many approaches for dealing with VANET attacks have been given in the past [5]. It is a network of geographically scattered autonomous vehicles that collaborate to deliver data to a centralized location known as a TMOC. Because the power required to transfer data between two vehicles is directly proportional to the distance between them, vehicles in a VANET send data to the TMOC in many hops to save power consumption and battery life. Because most VANETs are in remote locations where humans cannot reach them, and because batteries are difficult to replace once installed, designing a battery-powered network that is scalable and has a long life span becomes a challenge. Another problem with VANET is that it is a wireless network, because vital sensing data is delivered over wireless channels in the form of radio frequencies, network administrators must safeguard network data from attackers and intruders. One of the primary concerns in VANET is the implementation of various security measures such as authentication, encryption, and DDoS detection while keeping the limited power supply in mind [6].

VANETs operate in an IoV environment with little or no human intervention and the risk of vehicle capture; therefore vehicle performance is crucial. In order for a vehicle to detect an attack and take corrective action on its own [7–9] it must prevent the intruder from totally draining the battery. In general, assailants in the VANET can be divided into two types [10–12]:

(a) Exterior assailants/Outsider attackers: with little or no knowledge of the target;
(b) Interior assailants/Insider attackers: with limited or all knowledge of the target. Insider Assailants, who are genuine network members with access to all system secrets.

Outsider attackers attack the VANET in two ways: (a) passively by connecting to the appropriate frequency and obtaining all valuable information while causing no harm to the records; and (b) actively by connecting to the appropriate frequency and obtaining all valuable information while causing no harm to the data. Active attacks include DDoS, network jamming [13], collision attacks, replay attacks, and authentication attacks, among others. Insider attackers, on the other hand, can more severely attack the network because, as a legitimate user of the network, they can directly communicate with any reachable vehicle and can also transmit incorrect information in the network, resulting in severe attacks such as black hole and sink hole attacks.

A denial-of-service attack happens when an attacker floods a valid vehicle with continuous authentication packets or a continuous stream of data, leaving its services unavailable to other authorized vehicles as mentioned in Figure 3.1. Responding to authentication messages or receiving data packets from the attacker consumes all of the legitimate vehicle's resources, leading the battery to deplete fast [14]. DDoS is similar to DoS, except that it involves several attackers, resulting in a more severe attack. In this work,

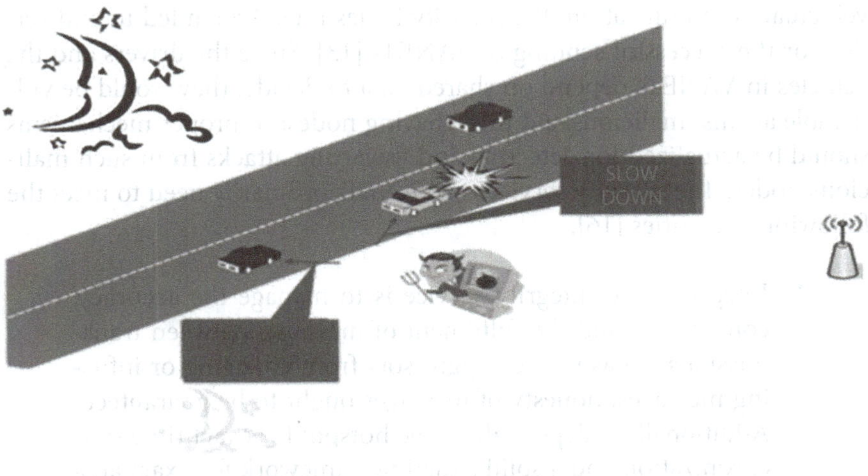

Figure 3.1 Assaults in VANET.

we used authentication and data filtration approaches to handle both inner and external DDoS attacks.

Specific Characteristics of VANET

High Dynamic Topology: VANETs have high effective topology. The correspondence interfaces between centre point changes rapidly. Correspondence between two nodes stays for less time. For instance, if two vehicles moving far from each other with a speed of 25m/sec and if the transmission run is around 250m, at that point the connection will keep going for just 5 seconds (250m/50ms-1). Thusly, this how uncommonly special topology is accessible in VANET.

Frequent disconnected Network: From the above specialty we can see that association between at least two vehicles stays for 5 seconds, or something like that. To keep up the continuous availability, vehicles need another connection close by instantly. Yet, in the event that failure happens, vehicles can interface with Road Side Unit (RSU). Frequently disconnected network principally happens where vehicle density is low, such as in a rural region.

Security Architecture

Requirements of Security Services

Security parts in MANETs have been broadly thought about; nevertheless, they are not sensible for VANETs due to the extraordinary properties of VANETs, so they can't be directly associated with VANETs. Notwithstanding an expansive scope of difficulties confronting securing vehicular communication, the security issues must be tended to and settled for the successful sending of VANETs [13]. Since the drivers and the vehicles in VANETs depend on shared data to decide, they would be vulnerable against malicious and misbehaving nodes; so proper mechanisms should be actualized for detecting and thwarting attacks from such malicious nodes. The security services of VANETs ordinarily need to meet the following necessities [16].

1. Integrity: The Integrity service is to manage the accuracy, consistency, and the fulfilment of messages between transmissions. So as to keep aggressors from changing or infusing messages, honesty of messages ought to be guaranteed. Additionally, a dependable time hotspot for exact time synchronization and a solid situating framework for exact area data could be utilized to ensure communication against attacks, for example, replay-attack or position mocking attack.
2. Availability: In VANETs, time basic messages, for example, emergency traffic data must be taken care of at any given time. In the event that one channel isn't accessible because of malicious behaviour or attack, there must be alternative intends to keep up vehicular network accessibility constantly.
3. Authentication: Each message traded must be validated to identify the sender of the message. Vehicles ought to respond just to data or occasions produced by genuine senders.
4. Non-repudiation: This administration is intended to identify misbehaving nodes or attackers and keep them from denying messages transmitted by them. Any vehicle-related data for communication, i.e., area, speed, and time, will be put away in a carefully designed On-Board Unit. It likewise could be utilized by specialists for examination to reproduce the scene of an accident with a similar arrangement and substance of the messages communicated before the accident.
5. Real-time constraints: Vehicles move at high speed. In a few circumstances like time-delicate communication, a continuous reaction is basic, so time limitations should be regarded.

Privacy: All driver data, for example, character, area and speed, should be secured against unapproved onlookers. Likewise, onlookers should not have the capacity to follow the routes of the vehicles.

Message Categories

Numerous applications are sitting waiting for organization in VANETs. These applications can be partitioned into two major categories, safety-related applications and non-safety-related applications [17].

1. Safety-related applications

   For example, collision shirking cautioning messages, emergency break cautioning messages, traffic light cautioning messages, or path merging cautioning messages could be sent to caution drivers [17, 18]. Since the messages sent by this sort of use enable drivers to settle on basic choices, guaranteeing the security and unwavering quality of such messages is essential.

   - Traffic information messages: Messages contain data, for example, street condition and accidents. Messages are sent to all vehicles inside particular zone for security and regularly they are not time critical messages.
   - General safety-related messages: This kind of messages is utilized for general safety applications, for example, helpful driving. Because of the high mobility of vehicle nodes, messages are time sensitive, so they ought to land inside the preset time window.
   - Liability-related messages: This kind of message is utilized for liability-related applications, which share traffic information, and drivers are in charge of the activity information. For example, if the message originator requires be followed back to research accident by the law authorization specialists, the experts ought to have the capacity to follow the message to its sender.

2. Non-safety-related Application

   There are non-safety-related applications, for example, activity enhancement, programmed installment administrations, area-related administrations, and driver infotainment administrations [17]. This sort of utilization does not have time-critical messages, yet securing messages for such administrations (e.g., installment benefits) and ensuring client protection (e.g., area benefit) are still critical for such applications.

3. Attacks on VANETs

   In VANETs, it is essential to represent non-participating substances in light of the fact that malicious nodes can purposely delude different vehicles by spreading false movement traffic information and degrade the system execution. In this segment, conceivable attacks in VANETs are discussed [18].

- Denial of Service Attack: In this sort of attack, the attackers endeavor to make the communication channels inaccessible or take control of a vehicle's assets. It can degrade the system's execution and furthermore influence the driver's security, particularly when safety-related application is influenced. For instance, if the attackers make massive system traffic on the road, when an accident happens, the vehicles moving into the danger area are kept from getting cautioning messages because of the denial of service attack.
- Message Suppression Attack: In this sort of attack, an attacker specifically drops messages from the system. The dropped message could be safety-related messages or basic data for the beneficiary. Additionally, the attacker may attempt to replay the dropped message later and misdirect the drivers.
- Fabrication Attack: In this kind of assault, an attacker attempts to transmit manufactured messages into the network. The messages sent by the attacker could contain false movement data or a fake identity message. It can likewise contain false warning messages and authentication.
- Alteration Attack: In this sort of attack, attackers attempt to modify existing messages. The attacker can change the substance of the message or postpone the message transmission.
- Replay Attack: In this sort of attack, the attacker attempts to send a prior message again to take advantage of the situation at the time of sending. Since a message is replayed, this is called replay attack.

Secure Routing and Data Dissemination in VANETs

The security part of the VANET framework is an imperative and has not been given the consideration it deserves [18]. As a result of the effect on the safety and security of travellers in the vehicles, planning protocols for delivering messages safely is critical.

Notwithstanding the communication device, numerous authors expect that every vehicle is furnished with a dependable situating device (e.g., a

Global Positioning System), so it can get precise area and time information [16]. To guarantee that all security requirements are precisely handled, we expect a situation where the adversaries can block any message in the VANET.

Due to their potential effect on the safety and security of individual in the vehicles, outlining protocols for trading messages safely is imperative. The security targets are authentication, non-disavowal of signalling packets, ensuring conditional user protection, recognizing and rectifying noxious information, and barring getting into malicious nodes from route discovery while messages are transmitted proficiently [16, 18].

3.2 Related Work

For peer-to-peer networks [13, 15], ad hoc networks [7, 13, 16], wireless sensor networks [17, 18], and the Internet of Vehicle [15], several reputation systems have been developed. These methods, however, cannot be applied to VANETs in their current forms since they do not take into account the key VANET properties of dynamicity and fast speed. The authors of [19] presented BTM, a beacon-based trust management system aimed at preventing internal attackers from delivering fraudulent messages in privacy-enhanced VANETs. In order to discern trustworthy event messages, a vehicle might employ not only direct or indirect event messages, but also beacon communications to build trust connections.

A "similarity-based trust management system for identifying bogus safety messages in VANETs" was proposed by Al Falasi and Mohamed [20]. Their trust system is built on resemblance. In VANETs, relationships are used to identify false safety event notifications from aberrant vehicles. Furthermore, it responds to a vehicle's claims of a safety occurrence and predicts that the source is the source. A true safety incident report will cause the car to react. A "trust-modelling framework for message propagation and assessment in VANETs" was suggested by Zhang *et al.* [21]. According to their model, a vehicle can determine whether or not to believe a message by weighing the opinions of others. However, in the extremely dynamic environment of a VANET, such decentralized trust mechanisms that rely on interactions with neighbors are impractical. TEAM (trust-extended authentication method) was suggested in [17]. It is a decentralized, lightweight authentication technique for highly dynamic VANETs that ensure integrity and non-repudiation, boosting the vehicles' communication confidence. TEAM, on the other hand, does not deal with the message data's dependability. The authors presented PTRS, a trust-based relay selection system based on Dirichlet distribution that differentiates the vehicles' trust levels while maintaining robustness, in [18]. Some attacks, such as packet analysis attacks, reputation link attacks, packet dumping assaults [22], and false reputation attacks, are resistant to the PTRS system. Das *et al.* [23] have presented strategies for determining a vehicle's trustworthy location. To begin, the information's trust percentage is calculated using the replies collected from automobiles. On this basis, the information's trust percentage is computed using the number of inquiries and the number of affirmative answers. Each car that responds positively to the information is awarded with points for delivering accurate information, allowing the network's trustworthiness to be calculated. Instead of travelling to an RSU or a TA in the second situation, when confidence is below 50%, the vehicle will assess the trustworthiness of each vehicle in the network. To identify malevolent vehicles and prevent them from being chosen as the GH, Mahmood *et al.* [24] devised a mixed trust management method. Their approach includes a composite measure for GH and proxy GH selection via intermittent elections (i.e., trust values ascribed to the vehicles along with their resource availability). This method aids in the formation of reliable and resource-efficient vehicular networks. Sugumar *et al.* suggested a trust-based security mechanism for VANETs based on clusters in [25]. The vehicles are grouped, and each vehicle's trust degree is calculated. The trust degree is made up of both direct and indirect trust degrees. Cluster heads are chosen based on the assessed trust degree. Messages are digitally signed by the sender's private key and encrypted using a public key (keys are provided by a trusted authority and decoded at the destination), and each vehicle is watched by a set of verifiers. This verifies both the sender's and receiver's identities, allowing the scheme to be authenticated. Hasrouny *et al.* [26] suggested a VANET Trust Model. To identify the most trustworthy vehicle as the GH, a combination of centralized and distributed collaboration between cars and infrastructure is used. This suggested approach uses a variety of measures to assess the behaviour of the cars in the group while guaranteeing participant privacy and minimizing network cost. To represent the local and worldwide trust linkages between vehicles, Hao *et al.* developed the notions of local trust and global trust. They employed the Page Rank algorithm [27], which is used to rank online sites, to calculate global vehicle trust [28].

In conclusion, the above-mentioned trust management methods are insufficiently dynamic to deal with the features of VANETs. Some of them [19] were suggested as solutions to various problems. Others [13, 17] relied on decentralized trust mechanisms, which are not practicable in the highly dynamic context of a VANET. A number of papers [7, 13, 15, 16, 24] attempted to address a specific sort of problem. It has a limited range of attack, which might be a drawback, since some others were built to tackle a wide range of situations. Authentication [19], privacy [18, 20], or localisation [21, 23] are only a few examples of specific challenges. As a result, we present, in contrast to prior methods, an adaptive trust-based security system for message transmission in VANETs based on VGA, our new clustering algorithm built to manage the dynamicity of such networks.

Ref. no.	Author and year of publication	Method name	Advantages	Disadvantages
[31]	Chung-Ming Haung, Tzu-Hua Lin, Kuan-Cheng Tseng, 2019	On Demand Member Centric Routing and Reactive Member Centric Routing	PDR and throughput are high in a platoon situation.	The cost of merged vehicles in RMCM is expensive.
[32]	Ahmed Nazar Hassan, Omprakash Kaiwartya, Abdul Hanan Abdullah, Dalya Khalid Sheet, Ram Shringar Raw, 2017	IVD based protocol for routing connectivity aware routing (IVD-CAR)	This technique outperforms CSR and A-CAR.	The impact of increasing traffic characteristics has not been examined.
[33]	Denghui Wang, Qingmiao Zhabg, Jian Liu, Dezhong Yao, 2019	Grid routing Protocol in wireless multimedia sensor network	When compared to previous methods, QOS is greater.	TGR protocol QOS is lower at first.
[34]	Na Lin, Yanjun Shi, Tongliang Zhang, Xuping Wang, 2019	Order aware hybrid genetic algorithm	The average margin is 42%, and the time is 580.72 seconds.	It is not up to the mark for CVRP algorithm.

(*Continued*)

(Continued)

Ref. no.	Author and year of publication	Method name	Advantages	Disadvantages
[35]	Iaonan Wang, Hongbin Cheng, Deguang Le, 2018	Intra VD and Inter VD routing algorithm for IPv4 based internet	Delay is decreased by 60%, while packet loss is reduced by 53%.	Changes are required when technology progresses.
[36]	Daxin Tian, K. Zheng, J. Zhoou, X. DUan, Y. Wang, Z. Sheng, Q. Ni, 2017	Unicast routing based on attractor selecting in VANET	Consciousness to discover the optimum answer	Path finding takes a long time to compute.
[37]	Menglan Hu, W. Liu, Kai Peng, X. Ma, W. Cheng, J. Liu, Bo Li, 2018	Vehicle assisted multi UAV routing and scheduling algorithm	Have lower time consumption and a lower distance expense.	More precision is required.

3.3 VANET Grouping Algorithm (VGA)

Trusted security scheme for message exchange in VANETs, is divided into two parts: (a) our VANET Grouping Algorithm (VGA), which is described below, to handle VANET characteristics such as dynamicity and high speed; and (b) a trust management scheme based on VGA. Pre-processing, setup, and maintenance are the three steps of VGA. Initial reputations are set in the pre-processing phase, while groups are generated in the setup phase. The maintenance phase is used to keep track of the vehicles' movement and to keep the created groups up to date. The reputation updating module, which is part of the trust management scheme, leverages the created groups as well as the authenticity of exchanged event messages to enhance or reduce vehicle reputations. The authenticity of event messages is calculated using three indicators: Lc (location proximity), Fc (number of forwarders), and Tc (time proximity). More information on these indicators may be found later in this section. According to [29, 30], a clustering algorithm's overall procedural flow is divided into four steps: neighbourhood discovery, cluster head selection, affiliation, announcement, and maintenance. As a result, VGA is built on comparable stages, with a few more to meet our specific requirements: management of trust.

Assumptions
The following assumptions underpin our algorithm:

- OBUs send out single-hop beacon signals including (at the very least) location, velocity, and direction on a regular basis.
- Each vehicle creates its own keys, makes a request for certificate signing to the TA, and receives a correctly signed certificate as verification.
- All of the vehicles' clocks are in sync.

3.4 Extension of Trusted Multipath Distance Vector Routing (TMDR-Ext)

Preventing both external and insider assaults is critical for preventing the bulk of VANET attacks. Authentication is the most effective means of preventing outsider assaults with the help of TMDR-Ext. Route request flooding is another common attack in VANET because attackers assault genuine vehicles by bombarding the network with authentication packets; in this scenario, traffic analysis becomes vital to prevent this attack. Insider attacks are very common in VANETs due to their distant deployment and low maintenance. Data flow analysis can also help to prevent insider attacks.

• Two-Way Validation
In this work, a mechanism is presented in which a vehicle may only connect with other vehicles in the network provided they know the network's common secret and protocol knows as TMDR-Ext routing protocol. The authentication system is based on a mathematical formula that is only known to valid network vehicles. To authenticate, each vehicle generates a random seed in TMDR-Ext, generates a question and response hash key based on that seed, and sends it to the other vehicles before they begin actual communication. When other vehicles correctly verify this secret information, they also authenticate themselves.

1. Vehicle A sends REQ (11) + RREQ to Vehicle B.
2. Vehicle B generates a random Clan and send [REP(22) + Clan (Y) + Hashing Key(KEY(Clan(Y)))] to Vehicle A.

3. Vehicle A validate Vehicle B by generating the Hashing Key at its own end and matching the same with the Hashing Key came from Vehicle B.
4. If Auth == Success then Vehicle A sends [REP(22) + Clan(X) + Hashing Key(KEY(Clan(X)))] to Vehicle B.
5. Vehicle B validates Vehicle A same as STEP(3).
6. If Auth == Success then Vehicle B updates its routing table and starts receiving information from Vehicle A.
7. If TIMER(Vehicle A) expires, Repeat STEP(2) to STEP(6).
   FUNCTION KEY(Clan) 1) KEY = SIN(Clan) * COS(Clan) * LOG10(Clan) 2) KEY = -KEY IF KEY < 0 3) KEY = ROUND(KEY*1000) 4) RETURN KEY

• Data Filtration Method in TMDR-Ext
Each vehicle in the network verifies the input data arriving from other vehicles in the network for traffic analysis in terms of multipath routing. Individual data quantities are then compared against threshold values to identify traffic anomalies. Because each vehicle in a VANET has a restricted data transfer rate, any vehicle detected with a data transmission range that exceeds that range is deemed an attacker vehicle, and all communication from that vehicle is stopped by all other vehicles in the network. The methodology of the preceding work is separated into three sections:

a) Algorithm Steps
b) Implementation
c) Results and Discussions

a) Steps of Algorithm

1. Vehicle A sends some data to Vehicle B.
2. If Vehicle A == validated Vehicle and Vehicle B == destination Vehicle, then Vehicle B check the information rate of the source Vehicle.
3. If information rate(s) > Threshold curb, then communication from that data source gets blocked.
4. Vehicle B informs the neighbor Vehicles of the source Vehicle about this attack.
5. If Vehicle A = validated Vehicle, Vehicle B checks its requesting rate or information rate, if it is exceeding the threshold curb, then Vehicle A gets blocked in this case.

6. Else Data from Vehicle A is accepted and forwarded to destination or the base station.

b) Implementation

1. Traffic generation using SUMO
2. Implementation using NS-2

Traffic generation using SUMO: SUMO was designed to replicate a city-sized traffic road network. The traffic flow is microscopically simulated. This implies that each vehicle moving inside the simulated network is represented separately and assigned a specific location (here we consider Vijay Nagar square with latitude 22.506894 and longitude 88.35444) and random speed as shown in Figure 3.2. These variables are changed every 1 second time step in response to the vehicles ahead and the roadway network the vehicle is travelling on. Street vehicle simulations are both time discrete and spatial continuous. The street features, such as maximum velocity and mobility models, are taken into account when simulating traffic.

Implementation using NS-2: Above algorithms have been implemented using ns2 simulators with implementation details and Vehicle communication details used under the simulation in which normal Vehicles send their data to the gateway Vehicles 10, 20, 30, 40 and 50, respectively. In this DDoS attack simulation both external and internal attackers try to bombard the packets either to the normal Vehicles or to the gateway Vehicles

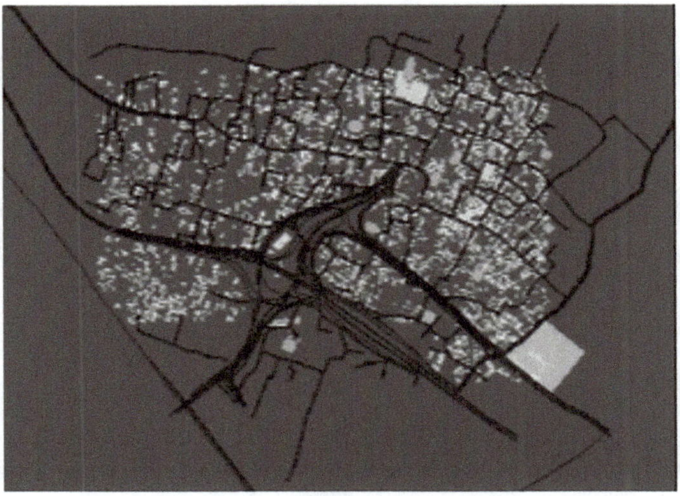

Figure 3.2 Vehicle movement area.

between 30 to 70 s. Vehicles which are attacked get busy in receiving the packets from the attacker and the normal functioning of the network gets degraded. Under the simulation three modes of operation have been maintained to check the network performance.

1. Initial actual Mode: When there is no attack and all the Vehicles work normally in the network using AODV routing.
2. Assailant Mode: When attackers attack the Vehicles and the network is unable to handle the situation using DDoS Routing.
3. Restraint Mode: When attackers attack the Vehicles and network has the ability to detect the attack and block the attacker Vehicles prevention of DDoS attack with the help of TMDR-Ext routing protocol.

This proposed TMDRExt routing protocol presents two techniques: one is a light-weight two-way authentication mechanism that will prevent most VANET assaults, and the other is a traffic analysis-based data filtering method that will identify and prevent VANET DDoS attacks. The Network Simulator 2 was used to assess several performance factors, including packet delivery ratio, throughput, and end-to-end delay with the help of NS2.

In Figure 3.3 vehicles are treated as vehicles, which contain attacker vehicle, Road Side Unit along with brink controller, normal vehicle and one trusted master of controller, which govern the communication in VANET.

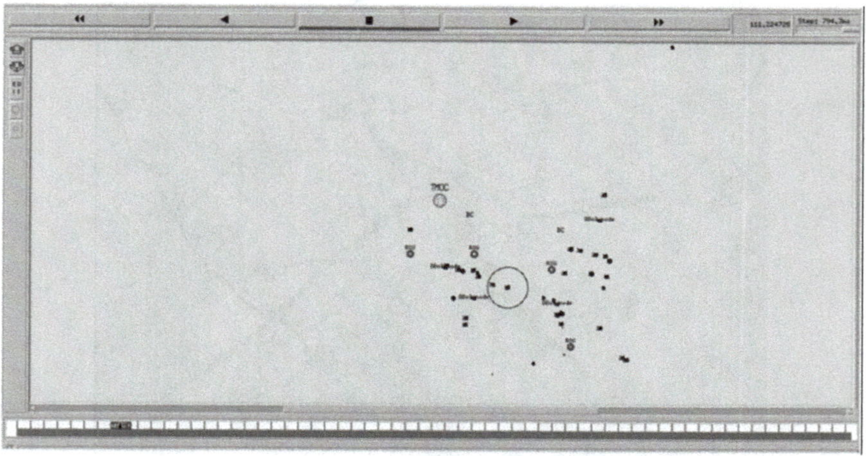

Figure 3.3 Simulation area.

Table 3.1 Simulation setup.

Simulation tool	**NS-2.35**
IEEE Scenario	802.11 Ext
Propagation	Two Ray Ground
Network area	10, 20, 30, 40, 50 vehicles
Traffic Type	UDP
Antenna	Omni-directional antenna
MAC Type	IEEE 802.11 Ext
Routing Protocol	AODV, DDos AODV, TMDR-Ext
Queue limit	50 Packets
Simulation area (in meter)	1000*1000
Queue type	Droptail
Channel	Wireless Channel
Simulation time	100 sec.

c) Results and analysis

In this work, we created a VANET scenario for NS-2 and then used several performance matrices such as packet delivery ratio, throughput, residual energy, and end-to-end latency to design different protocols. First, we developed a model file for IEEE 802.11Ext standard that must be used in conjunction with TCL script, then we constructed a TCL script that includes several routing protocols such as AODV, DDOS AODV, and TMDR-EXT, and last we constructed a VANET model or topologies. In the network, the network size is 2000m X 2000m, and the number of Vehicles is 10, 20, 30, 40, and 50. The following Table 3.1 the assumptions for this simulation:

❖Packet Delivery Ratio

The packet delivery ratio is the ratio of the number of packets originating by CBR source to the data packets received at the end destination by CBR sinks (Figure 3.4).

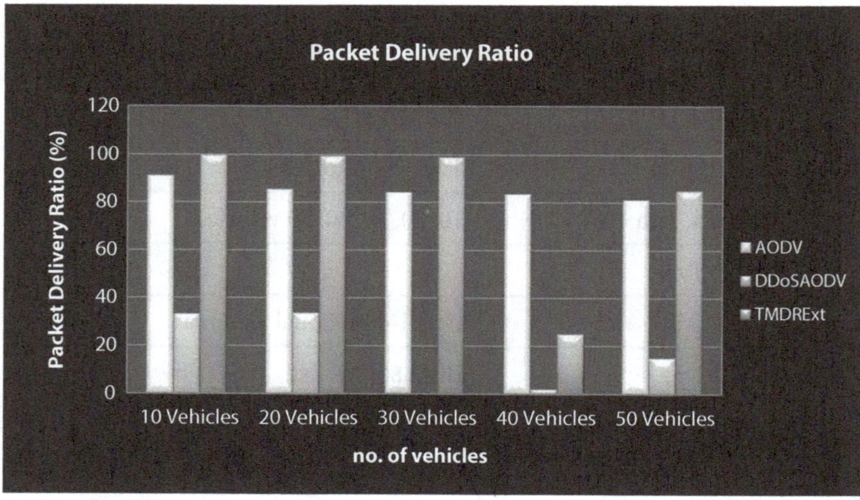

Figure 3.4 Packet delivery ratio.

❖Throughput

The protocol's throughput may be calculated as a percentage of data packets by the destination out of all packets sent by the source. The throughput is expressed in kilobits per second (kbps).

The Figure 3.5 sources were used to create this simulation process. In this paper, we look at a variety of characteristics in relation to varied pause periods.

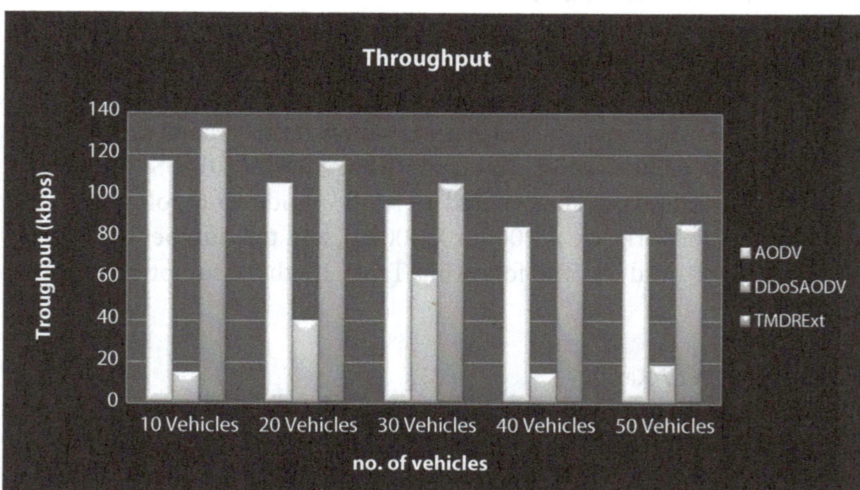

Figure 3.5 Throughput.

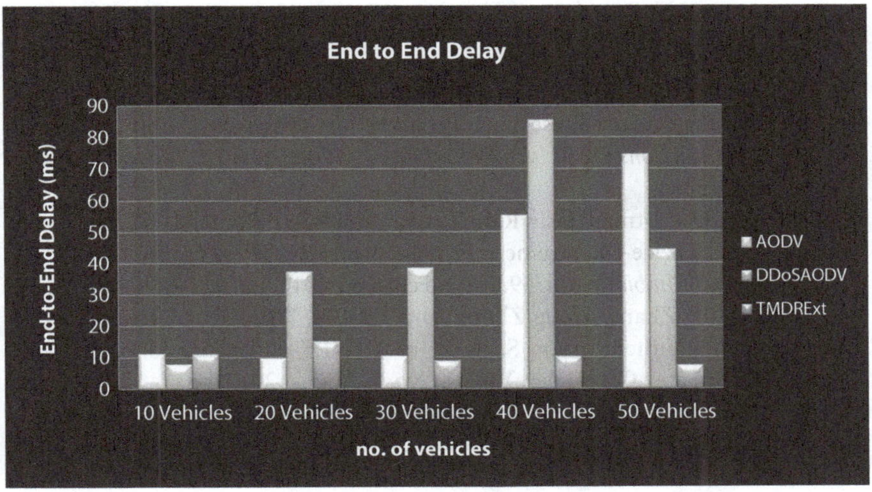

Figure 3.6 End-to-end delay.

❖Average End-to-End Delay

This includes delays caused by buffering of data packets during route discovery, queuing at the interface queue, retransmission delays at the MAC (Figure 3.6).

3.5 Conclusion

The Trusted Routing Technique was proposed, in which multiple Vehicle density-based circumstances are formed with the help of our suggested optimization. The use of the trust-based routing strategy in transferring data to the lead was studied and stressed, and the results demonstrate that employing the trust-based routing technique improves VANET efficiency. The trust-based routing approach may be used to develop a VANET routing protocol that is both efficient and secure. With a high quality of service awareness trustworthy approach, we present a trust-based routing strategy that employs the trust value mechanism to deliver data to the lead as evidence, resulting in fewer dropped packets and an extended network lifetime. When compared to alternative routing protocols such as AODV, the TMDR-EXT protocol offers a method to avoid DDoS attack utilisation in a VANET, according to the outcomes of this simulation.

References

1. Xiaodong Lin, Xiaoting Sun, Pin-Han Ho, and Xuemin Shen, "GSIS: A Secure and Privacy-Preserving Protocol for Vehicular Communications" *IEEE Transactions On Vehicular Technology*, Vol. 56, No. 6, November 2007, pp. 3442–3456.
2. Hanan Saleet, Otman Basir, Rami Langar, and Raouf Boutaba, "Region-Based Location-Service-Management Protocol for VANETs," *IEEE Transactions on Vehicular Technology*, vol. 59, no. 2, February 2010, pp. 917–931.
3. Jie Cui, Jing Zhang, Hong Zhong, and Yan Xu, "SPACF: A Secure Privacy-preserving Authentication Scheme for VANET with Cuckoo Filter," IEEE Journal of Latex Class Files, Vol. 14, No. 8, August 2015, pp. 1–14.
4. Pandi Vijayakumar, Maria Azees, Arputharaj Kannan, and Lazarus Jegatha Deborah, "Dual Authentication and Key Management Techniques for Secure Data Transmission in Vehicular Ad Hoc Networks," *IEEE Transactions on Intelligent Transportation Systems*, pp. 1–14.
5. Mithun Sahay Shrivastava, Ravi Khatri and Anand Singh Bisen, "Hybrid Approach for detecting and preventing DOS attack in VANET," *International Conference on Signal Processing, Communication, Power and Embedded System (SCOPES)-2016*, pp. 2021–2025.
6. Fengzhong Qu, Zhihui Wu, Fei-Yue Wang, and Woong Cho, "A Security and Privacy Review of Vanets," *IEEE Transactions on Intelligent Transportation Systems*, pp. 1–12.
7. M. Shahaya Sheela, A. Sivanantha Raja and V.R. Sarma Dhulipala, "Parametric Analysis of Mobile Ad-Hoc Network Environment," *International Journal of Computer Applications*, Volume 9, No. 9, November 2010, pp. 33–36.
8. Tingting Chen, Liehuang Zhu, Fan Wu, and Sheng Zhong, Stimulating cooperation in vehicular ad hoc networks: a coalitional game theoretic approach. *IEEE Transactions on Vehicular Technology*, 60(2):566–579, 2011.
9. David Chaum. Blind signature system. In *Advances in Cryptology*, pp. 153–153. Springer, 1984.
10. Monika, Sanjay Batish and Amardeep Singh, "Border-vehicle based Movement Aware Routing Protocol," *International Journal of Computer Science and Informatics*, ISSN Vol-1, Iss-4, 2012, pp. 2231–5292.
11. S. Yousefi, M. S. Mousavi and M. Fathy, "Vehicular Ad Hoc Networks (VANETs): Challenges and Perspectives," *6th International Conference on ITS Telecommunications*, 2006, pp. 761–766 (doi: 10.1109/ITST.2006.289012).
12. A. D. Devangavi and R. Gupta, "Routing protocols in VANET — A survey," *International Conference on Smart Technologies for Smart Nation (SmartTechCon)*, 2017, pp. 163–167, (doi: 10.1109/SmartTechCon.2017.8358362).
13. M. Shahaya Sheela, A. Sivanantha Raja and V.R. Sarma Dhulipala, "Parametric Analysis of Mobile Ad-Hoc Network Environment," *International Journal of Computer Applications*, Volume 9, No. 9, November 2010, pp. 33–36.

14. Munazza Shabbir, Muazzam A. Khan, Umair Shafiq Khan and Nazar A. Saqib, "Detection and Prevention of Distributed Denial of Service assaults in VANETs," *2016 International Conference on Computational Science and Computational Intelligence*, pp. 970–74.
15. Udit Agarwal and Monika Saxena, "Comparative and Behavioral Study of Various Routing Protocols in VANET," *International Journal of Advanced Research in Computer Science and Software Engineering* 3(10), October 2013, pp. 769–773.
16. Jinyuan Sun, Chi Zhang, Yanchao Zhang, and Yuguang Fang, "An Identity Based Security System for User Privacy in Vehicular Ad Hoc Networks" *IEEE Transactions on Parallel and Distributed Systems*, Vol. 21, No. 9, September 2010, pp. 1227–1239.
17. M. Chuang and J. Lee, "TEAM: Trust-Extended Authentication Mechanism for Vehicular Ad Hoc Networks," in *IEEE Systems Journal*, vol. 8, no. 3, pp. 749–758, Sept. 2014, doi: 10.1109/JSYST.2012.2231792.
18. Hao Hu, Rongxing Lu, "PTRS: A privacy-preserving trust-based relay selection scheme in VANETs", *Peer-to-Peer Networking and Applications* 10(5), September 2017, DOI:10.1007/s12083-016-0473-0.
19. Yi-Ming Chen and Yu-Chih Wei, "A Beacon-Based Trust Management System for Enhancing User Centric Location Privacy in VANETs", *Journal of Communications and Networks*, pp. 153–163, April 2013, DOI:10.1109/JCN.2013.000028
20. Hind Al Falasi, Nader Mohamed, "Similarity-Based Trust Management System for Detecting Fake Safety Messages in VANETs", *International Conference on Internet of Vehicles*, November 2015, DOI:10.1007/978-3-319-27293-1_24
21. Chen Chen, Jie Zhang, "A Trust Modeling Framework for Message Propagation and Evaluation in VANETs", *Information Technology Convergence and Services (ITCS)*, September 2010, DOI:10.1109/ITCS.2010.5581298
22. Alshammari, S.T.; Alsubhi, K., "Building a Reputation Assault Detector for Effective Trust Evaluation in a Cloud Services Environment." *Appl. Sci.* 2021, 11, 8496. https://doi.org/10.3390/ app11188496
23. P. Bagga, A. K. Das, S. Member, and Y. Park, "Authentication Protocols in Internet of Vehicles : Taxonomy , Analysis , and Challenges," *IEEE Access*, vol. 8, pp. 54314–54344, 2020, doi:10.1109/ACCESS.2020.2981397.
24. Siddiqui, S.A., Mahmood, A., Sheng, Q.Z., Suzuki, H., Ni, W., "A Survey of Trust Management in the Internet of Vehicles." Electronics 2021, DOI: https://doi.org/ 10.3390/electronics10182223
25. Sugumar, Dr & Alwar, Rengarajan & Jayakumar, C. (2018). Trust based authentication technique for cluster based vehicular ad hoc networks (VANET). *Wireless Networks* 24. 10.1007/s11276-016-1336-6.
26. Hasrouny, Hamsa & Samhat, Abed Ellatif & Bassil, Carole & Laouiti, Anis. (2018). Trust Model for Group Leader Selection in VANET. *International Journal of Digital Information and Wireless Communications*. 8. 10.17781/P002421.

27. C. Haydar, A. Roussanaly and A. Boyer, "Local Trust Versus Global Trust Networks in Subjective Logic," *2013 IEEE/WIC/ACM International Joint Conferences on Web Intelligence (WI) and Intelligent Agent Technologies (IAT)*, 2013, pp. 29–36, doi: 10.1109/WI-IAT.2013.5.
28. Tingting Chen, Liehuang Zhu, Fan Wu, and Sheng Zhong. Stimulating cooperation in vehicular ad hoc networks: a coalitional game theoretic approach. *IEEE Transactions on Vehicular Technology*, 60(2):566-579, 2011.
29. Cooper, Craig & Franklin, Daniel & Ros, Montserrat & Safaei, Farzad & Abolhasan, Mehran. (2016). *A Comparative Survey of VANET Clustering Techniques. IEEE Communications Surveys & Tutorials*. pp. 1–1. 10.1109/COMST.2016.2611524.
30. Xu, Dongkuan & Tian, Yingjie. (2015). A Comprehensive Survey of Clustering Algorithms. *Annals of Data Science*. 2. 10.1007/s40745-015-0040-1.
31. C. Huang, T. Lin, and K. Tseng, "Data Dissemination of Application Service by Using Member-Centric Routing Protocol in a Platoon of Internet of Vehicle (IoV)," *IEEE Access*, vol. 7 pp. 127713–127727, 2019, doi: 10.1109/ACCESS.2019.2936456.
32. A. N. Hassan and O. Kaiwartya, "Inter Vehicle Distance Based Connectivity Aware Routing in Vehicular Adhoc Networks," *Wirel. Pers. Commun* 98, 33–54 (2017), doi: 10.1007/s11277-017-4831-x.
33. D. Wang, "A Novel QoS-Awared Grid Routing Protocol in the Sensing Layer of Internet of Vehicles Based on Reinforcement Learning," *IEEE Access*, vol. 7, pp. 185730–185739, 2019, doi: 10.1109/ACCESS.2019.2961331.
34. N. A. Lin, S. Member, Y. Shi, and S. Member, "An Effective OrderAware Hybrid Genetic Algorithm for Capacitated Vehicle Routing Problems in Internet of Things," *IEEE Access*, vol. 7, pp. 86102–86114, 2019.
35. X. Wang, H. Cheng, and D. Le, "A routing scheme for connecting delay-sensitive urban vehicular networks to the IPv6-based internet," *Telecommun. Syst.* 69, 349–364 (2018), doi: 10.1007/s11235-018- 0443-3.
36. D. Tian, S. Member, K. Zheng, J. Zhou, X. Duan, and Y. Wang, "A Microbial Inspired Routing Protocol for VANETs," *IEEE Internet of Things Journal*, vol. 5, issue 4, pp. 2293–2303, pp. 1–10, 2017, doi: 10.1109/JIOT.2017.2737466.
37. M. Hu *et al.*, "Joint Routing and Scheduling for Vehicle-Assisted Multi-Drone Surveillance," *IEEE Internet of Things Journal*, vol. 6, issue 2, pp. 1781–1790, 2018, doi: 10.1109/JIOT.2018.2878602.

4

Technological Transformation of Middleware and Heuristic Approaches for Intelligent Transport System

Rajender Kumar[1*], Ravinder Khanna[2] and Surender Kumar[3]

[1]*Department of Computer Science and Engineering, Panipat Institute of Engineering and Technology, Panipat, India*

[2]*Department of Electronic Communication and Engineering, MM University, Sadopur, Ambala, Haryana, India*

[3]*Department of Computer Science and Engineering, Guru Teg Bahadur College, Punjab, India*

Abstract

There is a technological transformation from wireless telephony to VANET network, from the mobile node to a vehicle node and from the transport system to the Intelligent transport system. Based on the analysis of VANET network, middleware and heuristic approaches in problem-solving, it can be concluded that these topics are important influences to consider when designing and analyzing research problems. The review of literature played a critical role in discovering statistical knowledge about the middleware and heuristic approaches, gaps in published research and new original ideas for Intelligent transport.

Keywords: Vehicular networks, intelligent transport system, wireless sensor networks, mobile adhoc networks, controller area network, radio data system, transaction processing facility

4.1 Introduction

Wireless telecommunications refer to the transfer of information between two or more computers that are not physically connected. Distance can

**Corresponding author*: raj.mangyan@gmail.com

Romil Rawat, A. Mary Sowjanya, Syed Imran Patel, Varshali Jaiswal, Imran Khan and Allam Balaram (eds.) Autonomous Vehicles Volume 1: Using Machine Intelligence, (61–82) © 2023 Scrivener Publishing LLC

be short, such as a few meters for television remote control, or as far as thousands or even millions of kilometers for deep-space radios, cellular telephones, PDAs, and wireless networking [1]. Communication has gained huge growth in technology in multiple domains since its inception in 1880 [2]. Initially, wireless technology was used for personal communication. Now wireless technologies work at different frequencies to meet the requirements of corporate, personal and defense applications. There are various perspectives for wireless communications, including wireless sensor networks (WSN), mobile ad hoc networks (MANET), Wi-Max, and many others. The days of flying objects are currently noticeable in military applications, so that planes can communicate with each other in flight during a war or in similar cases. Wireless communication in vehicular objects on the road is commonly known as VANET [3].

The implementation of Vehicular Ad hoc Network (VANET) is widely adopted so that high-traffic regions can be controlled remotely with greater accuracy and performance [4]. VANETs employ mobile vehicles as network nodes. These can communicate between vehicles and/or roadside sensors. In this way, cars in the network are turned into wireless routers to connect among themselves and make a broad mobile network. The term VANET was first used in 2001 [5] under car-to-car ad-hoc mobile communication and networking applications. It was suggested [5] that VANET would provide navigation, road safety, and roadside services. It was proposed that VANET would be a crucial part of the Intelligent Transportation System (ITS) [6].

4.2 Evolution of VANET

VANETs are useful for V2V (vehicle-to-vehicle) communications in such a special network. Vehicles can also communicate with roadside infrastructure (V2I). These are two important, logical, and useful features of VANET. V2I enables the mobile node to connect to the Internet and enables global communication on the go. The detailed V2V function allows the exchange of data, for example, accidents and environmental conditions [7]. Figure 4.1 shows about the road side unit activity transformation relating to GPS and base station highlighting vehicular Ad hoc networks framework.

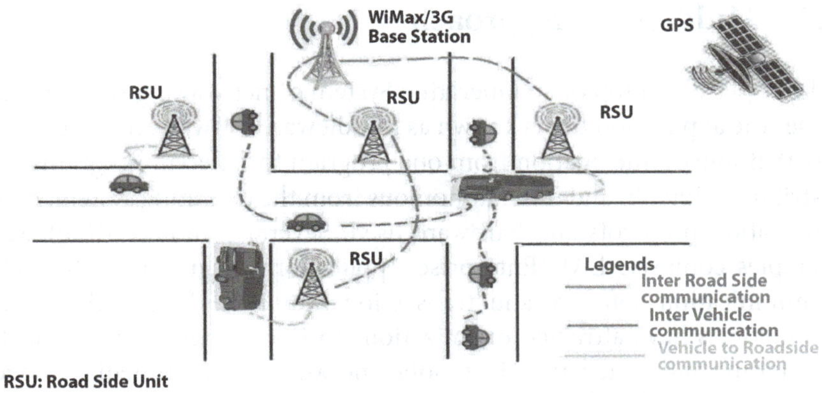

Figure 4.1 Vehicular Ad hoc networks.

The idea of utilizing radio communications in vehicles to enhance safety has been around for a long time, even before the appearance of digital communications. In recent years, vehicle safety has been focused on communication between vehicles. An example is the patent "Radio signalling systems for vehicles filed in 1922 and issued in 1925" [8]. In the mid-1980s, CAN (Controller Area Network), the earliest car controller network, was created by Bosch and is presently utilized in a few other automation applications [9].

In 1984, the Radio Data System Communication Protocol (RDS), which included limited quantities of digital information in radio frequency-modulated (FM) transmissions to transmit more audio signals over radio waves, turned into the main digital infrastructure for a vehicle (I2V), and a few years later it was announced in the United States as a Radio Data System. The Radio Data System (RDS) turned into the European standard in 1991.

Around 2005, after further testing, the RDS system was refined to offer the RDS-Traffic Message Channel (TMC). In the period 2000–2013, ISO, functioning with IEEE, built up a lot of standards for conducting these communication sessions with ETSI (European Telecommunications Standards Institute). Now that everyone has smartphones in every car, it is likely that all of these cell phones support the UHF (Ultra-High Frequency) wireless radio system, the Bluetooth, "Industrial, Scientific and Medical (ISM) in the 2.4 to 2.485 GHz band" in the fixed and mobile devices that help in building Personal Area Networks (PAN) [8].

4.3 Middleware Approach

A layer placed on top of an Operating System or networking software and under the application tier is known as Middleware software. It is a technology that moves information from one program to different programs in a distributed domain and it is autonomous from the operating system, communication protocols and hardware used. Several common Middleware examples comprise EAI (Enterprise Application Integration) tools, telecommunication software and transaction monitors. These Middleware technologies have attracted organizations to install them in some of the demanding and varied areas like mobile networking, multimedia, embedded systems and handheld devices [10].

In the early 1960s, IBM formed SABRE (Semi-automated Business Research Environment) for the airline system. This American system handles its aircraft reservations on centralized computer PCs. The original SABRE, though technically not a middleware system, gave the basics for the partition of transaction processing (TP), convey from equally to the operating system and the application. In the early 1970s, IBM improved upon the system and called it TPF (Transaction Processing Facility), which is an operating system designed for high volumes of transactions. These IBM-created systems with many improvements are still in use.

Middleware Fundamentals

Middleware like Java Virtual Machines and ACE (Armored Combat Earthmover) are functional to form reusable events, reusable event de-multiplexing, inter-process communication and object synchronization. Distribution middleware like CORBA (Common Object Request Broker Architecture) and Java RMI (Remote Method Invocation) from higher-level programming models are quality middleware that makes consistent and scalable applications.

Communication Support

Normally middleware uses distributed services over a network where applications interact throughout the network. Such distributed middleware is primarily of two types, (i) remote procedure called RPC and, (ii) messaging. The former uses the "client applications" to provide services to users. RPC hides the details of network programming from the application developers while the opposite approach is used by messaging, i.e., it exposes their presence to applications.

Concurrency Support

By concurrency support, middleware applications must require to be in terms of the numbers of messages processed in a given period. This period is essential for maximizing the concurrency in the system to enable the continuous operation of multiple tasks. Middleware systems use a different technique to improve concurrency, e.g., using threads inside a server process allows middleware communication systems to enhance supervision of network links and requests, and messages which show up on these connections. Execution and scalability are greatly dependent on the connection supervision and request dispatching facility, which the middleware delivers.

Common Middleware Services

Mainly middleware provides services to the network as follows: (i) Directory services: to discover resources dependent on their names, (ii) Transaction services: helps in rolling back the transactions which are used for reservations systems and financial applications, (iii) Security services: offers help for validation and approval over a distributed structure. Such facilities regularly organize over different unique security systems, (iv) Management services: observer and conserve systems while they are running, normally in manufacture, (v) Event services: grant applications to post occasion messages for a social affair to various applications. These can shape the purpose behind organization alerts, or for any expansive distribute/buy-in framework, (vi) Persistence services: help applications with taking care of their non-unpredictable information. There are loads of approaches giving diligence to applications. Some are worried about social information bases, others about article arranged data sets, (vii) Load adjusting services: adjusting the capacity of specific service replicas and visibly advancing individual request to the specific replica that is smallest stacked, (viii) Configuration services: enrich adaptability to applications by permitting middleware abilities to be altered. This sort of organization grants application lead, execution, and adaptability to be changed and changed without demanding the application to be recompiled. Sometimes, for instance, an application can have its security or worth-based limits engaged or incapacitated totally by methods for setup [10].

4.4 Heuristic Search

Heuristic technique, routinely called heuristic, is an approach to manage basic reasoning and learning, using a presence of mind system that isn't guaranteed as ideal or extraordinary, but is satisfactory to achieve fast destinations. In circumstances where finding the ideal course of action is impossible or strange, heuristic techniques can be used to quicken the route toward finding a reasonable plan [11]. One case of a decent broadly useful heuristic that is helpful for a group of combinational problems is the nearest neighbor heuristic, which works by choosing the locally superior option at each progression [12]. One of the best portrayals of the significance of heuristics in solving fascinating issues is How to Solve It [13].

Heuristic Problem Characteristics

Heuristic search is a general technique applicable to an enormous class of issues. It includes an assortment of explicit procedures, each one of which is especially powerful for a little class of issues. To pick the most fitting technique or blend of strategies for a specific issue, it is important to break down the issue alongside a few key measurements:

- Is the issue decomposable into a set of smaller or simpler subprograms?
- Can solution steps be overlooked or possibly fixed if they demonstrate impulsive?
- Is the issue's universe expectable?
- Is a decent answer for the issue clear without correlation with all other potential solutions?
- Is the ideal arrangement a condition of the world or a way to a state?
- Is a lot of information required to tackle the issue, or is information significant just to compel the search?
- Can a PC that is given the issue return the solution, or will the solution of the issue require a connection between the PC and an individual?

Heuristic Search Methods

1. Direct Heuristic Search Methods

   These are not commonly possible since they demand a great deal of time or memory. These methods look the entire state space for an answer and use an emotional mentioning of errands. Cases of these are:

   (a) Breadth-First Search is a procedure for searching tree or graph. It begins at the root of a tree or subjective node of a graph, occasionally stated a search key, and examines the totality of the neighbor nodes at the present depth before continuing ahead to the nodes at the subsequent depth.

It utilizes the opposite process as a depth-first search, which rather investigates the node branch beyond what many would consider possible before being compelled to backtrack and expand different nodes.

(b) Depth First Search is a calculation for exploring or glancing through tree or graphs. The computation starts at the root node and explores very far along each branch before backtracking. A variety of depth-first search was inspected in the "nineteenth century by French mathematician Charles Pierre Tremaux" as a technique for clarifying mazes [12].

2. Weak Heuristic Search Methods

These are also known as Informed Search, Heuristic Search, and Heuristic Control Strategy. These are fruitful at whatever point effectively use with right kinds of tasks and as a space explicit. This additional data is valuable to record preference amongst child nodes to examine. Individually node has a heuristic capacity related to it. Figure 4.2 represents about the heuristic technique improving the efficiency of a searching process.

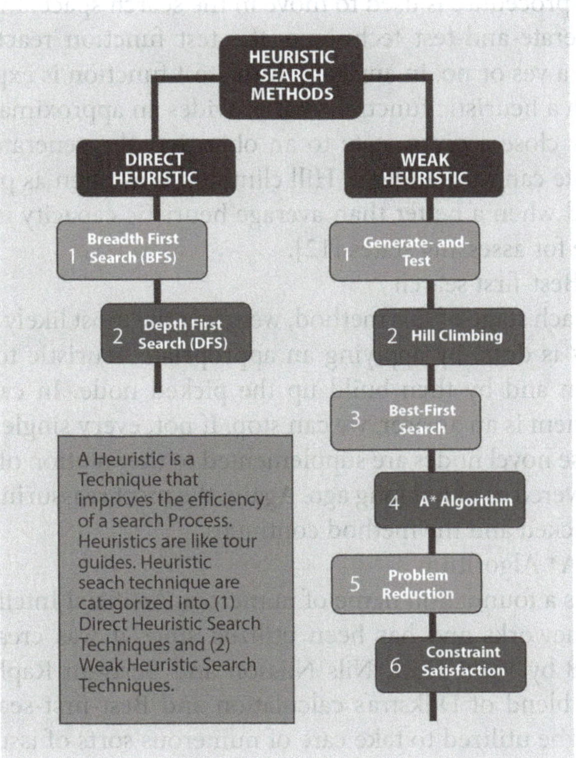

Figure 4.2 Heuristic search methods.

(a) Generate-and-test

The Generate-and-test system is the least difficult of the methodologies. Calculation of Generate-and-test is function as (1) Generate a possible result. For specific concerns, this suggests creating a particular point in a difficult space. For others, it infers creating away from a starting state. (2) Test to check whether this is an answer by isolating the picked point or the endpoint of the picked route to the plan of target states. (3) If an answer has been discovered, halted. Something different, return to step 1. The Generate-and-test is a depth-first inquiry as complete schedules need be created before they can be attempted. Generate and-test can, obviously, likewise work by producing solutions randomly; however, then there is no assurance that an answer will ever be found. The most direct approach to execute efficient Generate-and-test is as a depth-first search tree with backtracking [12].

(b) Hill climbing

It is a variety of generate-and-test in which reaction from the test procedure is used to move in the search space. In a pure generate-and-test technique, the test function reacts with just a yes or no. in any case, if the test function is expanded with a heuristic function that provides an approximation of how close a given state to an objective, the generate technique can mishandle it. Hill climbing is as often as possible used when a better than average heuristic capacity is available for assessing states [12].

(c) Best-first search

At each stage of this method, we select the most likely nodes. This is done by applying an appropriate heuristic to all of them and by then build up the picked node. In case one of them is an answer, we can stop. If not, every single one of those novel nodes are supplemented to the solution of nodes delivered as of not long ago. Again, the most reassuring node is picked and the method continues.

(d) A* Algorithm

A* is a foundation name of numerous Artificial Intelligence frameworks and has been utilized since it was created in 1968 by Peter Hart, Nils Nilsson and Bertram Raphael. It is a blend of Dijkstra's calculation and Best-first-search. It may be utilized to take care of numerous sorts of issues. A* search finds the most limited way through a search space to objective state utilizing heuristic function. A* requires heuristic function to assess the expense of path that goes through the specific state. It is characterized by the subsequent calculation.

$$f(n) = g(n) + h(n) \tag{4.1}$$

Where, g (n): The genuine cost path from the beginning state to the present status.

h (n): The genuine cost path from the present status to the objective state.

f (n): The real cost path from the beginning state to the objective state.

(e) Problem reduction

Problem reduction procedure utilizing AND-OR diagram is valuable for an issue that can be settled by disintegrating them into a set of minor issues which must be answered. Here, nodes whose successors should all be accomplished, and additionally nodes where one of the replacements must be accomplished (for example, they are options). This decay or decrease, creates arcs that we call AND circular segments. One AND arc may highlight various replacement nodes, which must all be understood all together for the arc to highlight an answer. As in OR diagram, a few arcs may arise out of a single node, demonstrating the assortment of manners by which the first issue may be tackled. That is the reason it is called AND-OR graph [12].

(f) Constraint satisfaction

Numerous issues in AI can be seen as issues of constraint satisfaction in which the objective is to find some issue express that fulfils a given arrangement of requirements. Instances of such an issue incorporate crypt-arithmetic puzzle and some genuine perceptual naming issues. Design tasks can also be seen as constraint-satisfaction problems in which a design must be made inside a fixed breaking point, on schedule, cost, and materials. Constraint satisfaction is a two-step process. Initially, constraints are found and circulated as far as throughout the system. In the second step, some hypothesis about an approach to support the constraints must be made [12].

Applications of Heuristic

(a) Algorithms Applying Artificial Intelligence

When a computer algorithm plays chess (for instance, Deep Blue), the computer cannot analyze each conceivable move. Rather, it will apply some practical rules to rapidly forego certain moves, concentrating on key advances that can lead to victory. When playing chess, master players can "see" a few moves forward. It is enticing to design an algorithm that could examine every conceivable move that players can have and research the effects of such a procedure on the result of the game. Be that as it may, such a calculation would need to investigate extremely numerous potential moves and would immediately turn out to be excessively moderate and requesting as far as memory resources to perform successfully. It is subsequently fundamental to utilize a heuristic method to rapidly dismiss certain moves which would in all likelihood lead to defeat while concentrating on moves that would appear to be a decent advance towards a success! We should think about the above situation when examining all potential moves for this white pawn. Could the computer settle on a fast choice concerning what might in all probability be the most ideal choice? Heuristic methodologies do not generally give you the best solution [11].

(b) Language recognition

By examining the estimation of the information entered by the client, for instance when the client poses an inquiry in the web crawler or interfaces with Google Home or the Amazon Echo gadget, the algorithm attempts to decide the information entered by the client. The heuristic algorithm can utilize word affiliations, setting investigation, past inquiry history, and flow internet searcher patterns/measurements to accelerate the pursuit procedure. Heuristic techniques can be utilized when creating calculations which attempt to comprehend what the client is stating, or requesting. For example, by searching for words affiliations, a calculation can limit the importance of words, particularly when a word can have two unique implications: for example, when utilizing Google search, a client types: "Raspberry Pi Hardware" one can deduce that for this situation, Raspberry has nothing to do with the organic product, so there is no compelling reason to give results on good dieting, cooking plans or markets [11].

(c) Big Data Analysis
This is needed when there is a lot of information to be examined. Web crawlers return indexed lists viably, profiling algorithms that the marketing division can use to decide target crowd individuals and their conduct, information investigation in logical exploration (for instance, medication to find circumstances-and-logical results connections between huge data sets) [11].

(d) Shortest Path Algorithms
GPS systems and self-directing vehicles additionally utilize a heuristic method to pick the best route from start to finish, for instance, the A* search algorithm. Further developed algorithms can likewise consider numerous elements, including types of road, speed limits, constant traffic information, etc., which can prompt very unpredictable algorithms. This is, for example, the case for the A* Search which proceeds concerning the distance as the crow flies from one node to second node to choose which way to investigate first and henceforth more viably locate the shortest path between two nodes. You can contrast two unique algorithms utilized which locate the shortest route from two nodes of a graph: 1) Dijkstra's Shortest Path Algorithm (Without utilizing a heuristic methodology), and 2) A* Search Algorithm (Using a heuristic methodology) [11].

(e) Machine Learning
Artificial intelligence algorithms dependent on machine learning, in which the computer makes an information base from past understanding, are another use of heuristic algorithms. For this instance, the algorithm utilizes decision and makes reasonable assumptions in light of past understanding. From time to time your algorithm needs to gather or examine information before applying a heuristic to make a decision.
This should be possible by executing an algorithm dependent on machine learning. For example, in a round of Top Trumps, if your algorithm knows the normal estimation of each card for every classification, it will have the option to choose the criteria which is in all probability going to win the round [11].

4.5 Reviews of Middleware Approaches

U. Lee, *et al.* proposed "MobEyes for urban observing dependent on the possibility of vehicle portability". The reproduction of a wrongdoing and, the back examination of an occasion checked by sensors, necessitate the gathering, loading and retrieval of enormous volumes of sensed data. The MobEyes exploited vehicles outfitted with camcorders and an assortment of sensors to perform occasion detecting, preparing, and message routing to different vehicles. The outcome demonstrated that MDHP "(MobEyes Diffusion/Harvesting Processor)" was versatile up to a huge number of nodes with restricted overhead and reasonable latency [14].

O. Riva, *et al.* introduced middleware that can care for mobile sensing applications in Urbanets. Urbanets aid a range of administrations going from crisis and surveillance, shopping, and entertainment. They utilized vehicles to give traffic data directions to make drivers aware of coming road jams and telephones. This architecture has been implemented in Java and tried on smartphones. Test results developed in specially Adhoc network of telephones have shown their plausibility with sensible exhibition as far as dormancy, memory, and vitality utilization [15].

Y. Bai, *et al.* proposed MidCASE "(Middleware Enabling Context-Awareness for Smart Environment)" in light of remote sensor networks. A service-oriented technique was adjusted to fulfil the needs of customizing scenarios in application development. The standard-based reasoning in the process raises a decent man-machine associating condition. They employed a healthcare scenario as a crucial function of middleware to validate the design principle of MidCASE [16].

L. Caviglione, *et al.* made a venture for building up an "Advanced Cooperative Infomobility System (ACIS)" to transform each moving vehicle into a kind of mobile sensor, able to watch and report about different occurrences identified with traffic. They executed a communique middleware for info motion applications on vehicular Adhoc systems. They outlined and inspired a portion of the structural decisions hidden in the ACIS middleware [17].

S. Nour, *et al.* introduced a middleware design for sensing and handling data produced in vehicular systems. This middleware provided the environment needed by software developers to create applications that could be used by the end users as well as other programmers. The research solution was suited for both large and small urban areas with the difference that in the first case the data gathered were more accurate and the problem of traffic congestion becomes more stringent [18].

M. Nitti, *et al.* investigated the fusion of "VANETs with the Social Internet of Things (SIoT), i.e., the Social Internet of Vehicles (SIoV)". In the SIoV each vehicle was fit for setting up social associations with different vehicles and the roadside units (RSUs). Furthermore, the authors introduced a simulation examining sensible vehicular mobility to consider the qualities of the subsequent social organization structure [19].

M. Zennaro, *et al.* introduced a model for multi-vehicle applications, fitting for dynamic heterogeneous dispersed applications, specifically, those including multi-vehicle activity situations. They built up a middleware dependent on Java and JavaSpaces innovation, and a calculation, for naturally assembling and reconfiguring administration systems. The Transaction Manager administration was utilized by the JavaSpace and by sweepers to keep up information consistency. They outlined the methodology with a multi-vehicle search-strategic [20].

Chien-Chung Shen *et al.* presented middleware design that encourages questioning and observing of sensor systems. "A sensor execution environment (SEE)", running on every sensor node, was liable for forwarding incoming messages and inspects entirely appearance "SQTL (Sensor Query and Tasking Language) messages". They found that the quantity of valuable detecting from the two calculations were the very same (209) while the common algorithm performed more sensing exercises "(76,800 times compared to 8,828 times when using the coordinated algorithm)" [21].

R. Meier, *et al.* presented the execution of "STEAM (Scalable Timed Events and Mobility)", an event-centered middleware for shared, area-specific applications utilizing wireless LAN. STEAM contrasts from other event-centered middleware in that its design does not depend on the nearness of any architecture, occasion warning filtering might be circulated and might be applied to attributes of event notification. STEAM has been intended for IEEE 802.11b-based network. The authors' assessment of STEAM demonstrated that utilizing proximity-based filtering to guarantee the propagation scope of event warnings minimizes the quantity of event conveyed to an application [22].

R. Meier, *et al.* structured an "occasion-based middleware, called RT-STREAM", which depended on an implicit-event model and had been utilized for versatile applications also especially vehicular systems. RT-STREAM varies from other networks in that it does not depend on the nearness of any different infrastructure. RT-STEAM provisions dispersed methodologies for finding peers and routing events utilizing a distributed addressing. They combined communication features such as proactive, mobility and resource reservation protocol with a time-bounded system using TBMAC protocol to accomplish more results [23].

W. Zhaohui, *et al.* built up a versatile middleware platform ScudWare for smart vehicle space. In this, strategies of multi-agent, context-aware, and versatile module boards were easily combined. They evaluated the ScudWare with a testbed of smart vehicle space in terms of, a) ScudWare extension; b) coordination of the biological confirmation and AI advances; and 3) improvement applications for ITS in the smart vehicle utilizing ScudWare stage [24].

E.P. De Freitas, *et al.* introduced a middleware architecture that permits the combination and collaboration among WSN nodes and little "Unmanned Aerial Vehicles (UAVs)" actualizing an observation system. They performed simulations utilizing Shox, Java-based wireless network simulator for the aggregate of 20 runs. The arrangement displayed a system of "8000 nodes randomly circulated on a region with measurements 5Km x 5Km, in which 2000 nodes" had the sensing capacities. They performed the simulation with a sole and tiny agent that fits in a communication packet of IEEE 802.11b [25].

M.O. Killijian, *et al.* introduced a middleware design for mobile applications that gives a new structure dependent on cooperative methodologies like proximity map, cooperation oracle and data backup service. The research work concerned about the arrangement of a critical data backup, a trust and cooperation oracle, and a proximity map. They depicted another platform for the test assessment of mobile application and indicated that the mobility aspects of the platform were improved [26].

G.S. Blair, *et al.* introduced the CONNECT middleware architecture as new middleware that progressively created a distributed architecture foundation for the current working environment. The authors concentrated on the key part of middleware, in particular, the job of ontologies in supporting centre fundamental middleware capacities identified with accomplishing interoperability. They made progressively proper availability arrangements as far as understanding a given deployment environment [27].

R. Beaubrun, *et al.* recommended a middleware architecture for dispersing delay-obliged data in WSNs. The authors deployed the middleware in three fundamental devices: (1) the sensor nodes, (2) the gateway and (3) the base station to assess a genuine situation. They performed 20 investigations, sending 400 notifications to the clients all through three protocols: (1) SMS, (2) email and (3) Twitter. In light of this data, two unique examinations were played out: the investigation of singular messages and the examination of the experiment. They understood that the max delay never surpasses the pre-established threshold [28].

J. Luo, *et al.* proposed an "SOA-based middleware (VsdOSGi)" for vehicular architecture, which enables a vehicle to impart services to others. They structured an algorithm dependent on QoS (SDQ) to guarantee offering ideal service. They demonstrated that middleware can viably decrease the quantity of message sending times, minimize bandwidth capacity while improving network use and service discovery rate with less response time [29].

T.A. Butt, *et al.* proposed a versatile SIoV (Social Internet of Vehicles), dependent on web technology to give an establishment to creating SIoV applications, and stress the significance of web technology to come across the necessary interoperability for supporting the structure of different services. Authors featured the social connections between various elements of the architecture alongside the administration of these connections remembering the dynamic nature of SIoV architectures [30].

Y. Zheng, *et al.* proposed a "context-based location privacy protection middleware architecture, named PP-OSGi" for simplified application expansion by protecting the heterogeneity of automobile-detected information. Besides, to secure the genuine location data of request vehicles underneath various vehicle densities in a specific region, the authors proposed a "dynamically adjustable k-anonymous (DAK) algorithm" and a "location privacy protection (DLP) algorithm" dependent on a spurious area, which was completely captured in PP-OSGi with improved service request success rate [31].

4.6 Reviews of Heuristic Approaches

H. Saajid, *et al.* analyzed V2V and V2R interchanges utilizing the heuristic methodology considered short-range protocols: ZigBee, Wi-Fi, and Bluetooth with information rates, time, and range; different parameters included roadway, urban streets, and blended. It worked out that, without RSUs, Bluetooth could not move any messages in light of its little range, the quality of ZigBee extended between 5% and 38%, and the Wi-Fi went somewhere in the range of 30% and 90% [32].

P.K. Singh, *et al.* contributed an inclusive study of vehicular communication. They introduced the particulars of the 5G network, its range, use cases, designs and status. They featured the regulation efforts of IEEE, ISO, ITU, ETSI, etc. They additionally talked about ITS protocol stacks as an associated vehicle, "ARIB STD-T109 and Cooperative-ITS in the USA, Japan, and Europe". They introduced the rising radio access advances, for example, consistent availability, edge, fog, programming characterized and data network [33].

R. Saxena, *et al.* approached the Travelling Salesman Problem employing GA method which more than once alters a population of distinct results, with the additional power of present-day processing systems. They thought of an equal kind of GA for both multicore and many-core designs over Open Multi-Processing and CUDA (Compute Unified Device Architecture) to make challenging issues like "Vehicle Routing Problem for Google maps and DNA sequencing". They investigated both CUDA and OpenMP for different degree and structure of graphs portraying adjacent estimate to the exactness as far as the most ideal path for traversal with extremely weakened execution time [34].

W. Zhang, *et al.* devoted this exceptional issue on heuristic pursuit in AI to Claude Shannon and Herbert Simon. Heuristic pursuit stays as a centre zone of "artificial intelligence (AI)". They included ten articles that right now were genuinely illustrative of research on search and heuristics. Bonet and Geffner built up a group of heuristic inquiry organizers dependent on a basic domain-independent heuristic, and showed that the new organizers are serious with probably the best present organizers. Hansen and Zilberstein thought about scanning for optimal policies to take care of issues. Markov choice prepared, a typical system utilized in AI for arranging and learning under vulnerability [35].

M. Fathian, *et al.* presented novel clustering algorithms based on "Ant Colony System (ACS)" and "Artificial Immun System (AIS)" as metaheuristic algorithms. They presented results of the proposed algorithms and compared them with six VANET clustering algorithms from the literature. The results showed that the ACS-BASED algorithm and AIS-based clustering algorithm performed the same as the Lowest-ID algorithms and Highest-Degree algorithms, respectively [36].

A. Amuthan, *et al.* explored the basic concepts in Tabu Search (TS) in application areas like VANET. They applied the metaheuristic approach Tabu Search to find the optimized solution in search process with minimum cost function. The results focused on the benefits of integrating Tabu with other heuristics algorithm like Genetic Algorithm to give optimum solution to find neighbors for disseminating data [37].

K. Jyothi, *et al.* proposed a unique algorithm named "Ant Queue Optimization (AQO)" which was an expansion of "Ant colony optimization algorithm (ACO)". They indicated that the AQO has occupied a smaller time to progression all the vehicles inside a specified organization when contrasted with ACO. A few examinations had been completed to contrast ACO and AQO through a set of vehicles, i.e., 10, 30, 50, 70, 100, 200. They indicated that the AQO has occupied a smaller amount of time to progression all the vehicles inside a specified organization when contrasted with ACO [38].

S. Dornbush, *et al.* proposed a framework that utilizes a standard GPS driving guide, increased with shared wireless communication. The framework utilized a blend of clustering and epidemic communication to discover and spread unique traffic designs. They acquainted a strategy with discover clusters in a detached distributed system. The framework performed well in testing conditions with non-stationary information and a sparsely associated dynamic system, intended to give high esteem data to drivers in an auspicious manner [39].

W. Zhigang, *et al.* proposed another clustering strategy for enormous multi-hop vehicular Adhoc systems. This research introduced the solidness of the projected cluster assembly and communication expense for keeping up the construction and connectivity in an application setting. They executed CORSIM and NS-2 simulators with genuine traffic by simulating a network of 100 cars moving in an expressway with 10 km length. The packet size containing the positional data was 75 bytes and the traffic thickness was 50 vehicles/km per track. The simulation demonstrated that strategy was adaptable and stable [40].

E. Vanet, *et al.* introduced two heuristic strategies for voltage constraint management through LV (Low Voltage) planned as a "DSO (Distribution System Operators)" subordinate help. In light of voltage sensitivity investigation, these techniques expected to assist the DSO with benefitting the LV adaptability accessible on a given LV network to settle voltage deviations. A heuristic strategy was additionally introduced to refine the solution and permit the best LV adaptability dispatch utilizing low computational need [41].

L. Urquiza-Aguiar, *et al.* proposed "a generic Geographical Heuristic Routing (GHR)" procedure that can be functional to any geographical routing protocol that creates sending results hop-by-hop. GHR comprised Tabu-search meta-heuristics in its operation to improve local search results. The results indicated the utilization of a Tabu list rundown adds to refining the packet conveyance proportion by nearby 5% to 10% [42].

K. Ravikumar, *et al.* utilized heuristic techniques to characterize heuristic procedures and discover a possible and suitable solution. They considered the bottleneck control issues in VANET, and arranged one congestion controller procedure dependent on Tabu Search calculation. The outcomes demonstrated that tuning transmission range and transmission rate utilizing Tabu search can minimize delay and extent of packet damages. Just as, they concluded that in any event 3.86% was upgraded the amount inside

VANETs when control mobbing utilizing Tabu Search rather than other congestion control techniques [43].

R. Saxena, *et al.* discussed a hybrid approach genetic algorithm centered heuristic for mobicast routing, to tackle the issue of packet drops because of flooding or broadcasting. The achievement of such heuristic centered routing algorithm for MANETs opens the entryway of consideration for VANETs. They improved the proposed strategy utilizing similar processing command of multi-core and many-core processors utilizing "OpenMP and Computationally Unified Device Architecture (CUDA) API" [44].

S.P. Sasirekha, *et al.* structured a heuristic for adequate transmission in the VANET condition. The performance of heuristic was effective for packet transmission, and to decrease the system overhead. They involved "scatter message and remain idle until receiving the response (SM-RIRR)" approach for (RMACO) to help with assessing the reaction time [45].

O. Senouci, *et al.* suggested a novel "Heuristic one-hop Clustering Algorithm based on RSU (HCAR)", for IoV. Also, this effort minimized the probabilities of the inaccessibility of CH by choosing a Secondary CH (SCH), and to respond to repeated variations in the network topology. They led simulations utilizing the NS2 and the VanetMobiSim coordinated condition. The scheme outperformed "Mobility aware and SIngle-hop Clustering scheme MOSIC" and Distributed Multi-Hop Clustering algorithm for VANETs [46].

4.7 Conclusion and Future Scope

In VANETs, the fusion of middleware and heuristic approaches played an important role. However, Intelligent Transport System of the next generation would need high mobility, low latency, real-time applications and networking autonomous cars, which cannot be solved by traditional computing. Therefore, combining these approaches with computing standards provided a possible solution to different issues. Middleware consists of Advanced Wireless Modules, Effective Algorithms, Database servers, software servers, content management systems, etc., tools that facilitate the creation and implementation of product. Nature-Based Algorithms is a highly active field of study which can be integrated for future scope. Many meta-heuristic-based works explain scientific observations focused on computer-driven algorithmic studies. Meta-heuristic methods can be extended to a variety of problems separate from the issue.

References

1. R. Kumar, R. Khanna, S. Jangra, and P. K. Verma, "A study of diverse wireless networks," *IOSR Journal of Engineering*, vol. 2, no. 11, pp. 01–05, 2012.
2. R. Kumar, R. Khanna, and S. Kumar, "Deep learning Integrated approach for collision avoidance in Internet of Things based smart vehicular networks," *Journal of Advanced Research in Dynamical and Control Systems*, vol. 10, no. 14, pp. 1508–1512, 2018.
3. T. Camp, J. Boleng, and V. Davies, "A survey of mobility models for ad hoc network research," *Wireless Communications and Mobile Computing*, vol. 2, no. 5, pp. 483–502, 2002.
4. R. Kumar, R. Khanna, and S. Kumar, "An effective framework for security and performance in Intelligent Vehicular ad-hoc network," *Journal of Advanced Research in Dynamical and Control System*, vol. 10, no. 14, pp. 1504–1507, 2018.
5. C.K. Toh, *Ad Hoc Mobile Wireless Networks: Protocols and Systems*, Prentice-Hall, 2001.
6. M. Kaur, S. Kaur, and G. Singh, "Vehicular ad hoc networks," *Journal of Global Research in Computer Science*, vol. 3, no. 3, pp. 61–64, 2012.
7. M.NMajeed, S.P Chattha, A. Akram, and M. Zafrullah, "Vehicular Adhoc networks history and future development arenas," *Information Technology & Electrical Engineering Journal*, vol. 3, no. 1, pp. 1–5, 2013.
8. M. Annoni, and B. Williams, "Vehicular Adhoc networks - standards, solutions, and research," Springer International Publishing, Switzerland, pp. 3–22, 2015.
9. A.M.Vegni, M. Biagi, and R. Cusani, "Smart vehicles, technologies and main applications in vehicular ad hoc networks," doi 10.5772/55492, pp. 3–20, 2013.
10. S. Vinoski, "An Overview of Middleware," *International Conference on Reliable Software Technologies*, pp. 35–51, 2004.
11. D.W. Patterson. *Introduction to Artificial Intelligence and Expert Systems*, 1st ed. Pearson Education India, 2015.
12. E. Rich, and K. Knight. *Artificial Intelligence*, 2nd ed. Tata McGraw-Hill Publishing Company Limited, New Delhi, 2002.
13. G. Polya. *How to solve it: A new aspect of mathematical method,* 2nd ed. Doubleday anchor books, Doubleday & Company, Inc., Garden City, New York, 1957.
14. U. Lee, B. Zhou, M. Gerla, E. Magistretti, P. Bellavista, and A. Corradi, "MobEyes: smart mobs for urban monitoring with a vehicular sensor network," *IEEE Wireless Communications*, vol. 13, no. 5, doi:10.1109/WC-m.2006.250358, pp. 52–57, 2006.
15. O. Riva, "Middleware for Mobile Sensing Applications in Urban Environments," Academic Dissertation, University of Helsinki, Finland, 2007.

16. Y. Bai, H. Ji, Q. Han, J. Huang, and Z. Zhang, "Towards a service-oriented Middleware Enabling context awareness for smart environment," *International Journal ad Hoc and Ubiquitous Computing*, Inderscience Enterprise Limited, vol. 4, no. 1, pp. 24–35, 2009.
17. L. Caviglione, G. Ciaccio, and V. Gianuzzi, "Architecture of a communication middleware for VANET applications," 2011 *The 10th IFIP Annual Mediterranean Ad Hoc Networking Workshop*, doi:10.1109/med-hoc-net.2011.5970474, IEEE, pp. 111–114, 2011.
18. S. Nour, R. Negru, F. Xhafa, F. Pop, C. Dobre, and V. Cristea, "Middleware for data sensing and processing in VANETs," 2011 *International Conference on Emerging Intelligent Data and Web Technologies*, 978-0-7695-4456-4, IEEE Computer Society, pp. 42–48, 2011.
19. M. Nitti, R. Girau, A. Floris, and L. Atzori, L, "On adding the social dimension to the Internet of Vehicles: Friendship and middleware," *2014 IEEE International Black Sea Conference on Communications and Networking* (BlackSeaCom). doi:10.1109/blackseacom.2014.6849025, pp. 134–138, 2014.
20. M. Zennaro, J. Ko, R. Sengupta, and S. Tripakis, "A Service Network Architecture for a Multi-Vehicle Search Mission," Decision and Control, *Proceedings of the 40th IEEE Conference*, Vol. 2, pp. 1503–1508, 2001.
21. S. Chien-Chung, C. Srisathapornphat, and C. Jaikaeo, "Sensor information networking architecture and applications," *IEEE Personal Communications*, vol. 8, no. (4), pp. 52–59, 2001.
22. R. Meier, and V. Cahill, "Exploiting Proximity in Event-Based Middleware for Collaborative Mobile Applications," *Lecture Notes in Computer Science*, doi:10.1007/978-3-540-40010-3_26, pp. 285–296, 2003.
23. R. Meier, B. Hughes, R. Cunningham, and V. Cahill, "Towards Real-Time Middleware for Applications of Vehicular Ad Hoc Networks," *Lecture Notes in Computer Science*, doi:10.1007/11498094_1, pp. 1–13, 2005.
24. Z. Wu, Q. Wu, H. Cheng, G. Pan, M. Zhao, and J. Sun, "ScudWare: A semantic and adaptive middleware platform for smart vehicle space," *IEEE Transactions on Intelligent Transportation Systems*, vol. 8, no. 1, pp. 121–132, 2007.
25. E.P. De Freitas, A.M. Ferreia, C.E. Pereira, and T. Larsson, "Middleware Support in Unmanned Aerial Vehicles and Wireless Sensor Networks for Surveillance Applications," *Intelligent Distributed Computing III*, doi:10.1007/978-3-642-03214-1_33, pp. 289–296, 2009.
26. M. Killijian, M. Roy, and G. Severac, "ARUM: A cooperative middleware and an experimentation platform for mobile systems," *2010 IEEE 6th International Conference on Wireless and Mobile Computing, Networking and Communications*, doi:10.1109/wimob.2010.5645030, pp. 442–449, 2010.
27. G.S. Blair, A. Bennaceur, N. Georgantas, P. Grace, V. Issarny, V. Nundloll, and M. Paolucci, "The Role of Ontologies in Emergent Middleware: Supporting Interoperability in Complex Distributed Systems," *Lecture Notes in Computer Science*, doi:10.1007/978-3-642-25821-3_21, pp. 410–430, 2011.

28. R. Beaubrun, J.F. Llano-Ruiz, B. Poirier, and A. Quintero, "A middleware architecture for disseminating delay-constrained information in wireless sensor networks," *Journal of Network and Computer Applications*, vol. 35, no. 1, doi:10.1016/j.jnca.2011.09.002, pp. 403–411, 2012.
29. J. Luo, T. Zhong, and X. Jin, "Service discovery middleware based on QoS in Vanet," *12th International Conference on Natural Computation, Fuzzy Systems and Knowledge Discovery IEEE*, pp. 2075–2080, 2016.
30. T.A. Butt, R. Iqbal, S.C. Shah and T. Umar, "Social Internet of Vehicles: Architecture and enabling technologies," *Computers and Electrical Engineering*, ELSEVIER, vol. 69, no. 2018, pp. 68–84, 2018.
31. Y. Zheng, J. Luo and T. Zhong, "Service recommendation middleware based on location privacy protection in VANET," *IEEE Access*, vol. 8, pp. 12768–12786, 2020.
32. H. Saajid, W. Di, W. Xin, S. Memon, N.K. Bux, and Y. Aljeroudi, "Reliability and Connectivity Analysis of Vehicular Ad Hoc Networks under Various Protocols Using a Simple Heuristic Approach," *IEEE Access*, pp. 1–11, 2019.
33. P.K. Singh, S.K. Nandi, and S. Nandi, "A tutorial survey on vehicular communication state of the art, and future research directions," *Vehicular Communications*, vol. 18, no. 2019, pp. 1–39, 2019.
34. R. Saxena, M. Jain, S. Bhadri, and S. Khemka, "Parallelizing GA based heuristic approach for TSP over CUDA and OPENMP," *International Conference on Advances in Computing, Communications and Informatics (ICACCI)*, pp. 1934–1940, 2017.
35. W. Zhang, R. Dechter, and R.E. Korf, "Heuristic search in artificial intelligence," *Artificial Intelligence*, vol. 129, no. 2001, Elsevier, pp. 1–4, 2001.
36. M. Fathian, A.R. Jafarian-Moghaddam, and M. Yaghini, "Improving vehicular Adhoc network stability using meta-heuristic algorithms," *International Journal of Automotive Engineering*, vol. 4, no. 4, pp. 891–901, 2004.
37. A. Amuthan, and K.D. Thilak, "Survey on Tabu Search Meta-Heuristic Optimization," International Conference on Signal and Processing, Communication, Power and Embedded System (SCOPES) -2016, pp. 1539–1543, 2016.
38. K. Jyothi, and J.C. Jackson, "A time-based approach for solving the dynamic path problem in VANETs – An extension of Ant Colony Optimization," *Journal of Engineering Science and Technology*, vol. 13, no. 3, pp. 813–821, 2018.
39. S. Dornbush, and A. Joshi, "StreetSmart Traffic: Discovering and Disseminating Automobile Congestion Using VANET's," *IEEE 65th Vehicular Technology Conference* - VTC2007-Spring, pp. 11–15, 2007.
40. W. Zhigang, L. Lichuan, Z. MengChu Zhou, and N. Ansari, "A Position-Based Clustering Technique for Ad Hoc Inter-Vehicle Communication," *IEEE Transactions on Systems, Man, and Cybernetics, Part C (Applications and Reviews)*, vol. 38, no. 2, doi:10.1109/tsmcc.2007.913917, pp. 201–208, 2008.

41. E. Vanet, S. Toure, N. Kechagia, R. Caire, and N. HadjSaid, "Sensitivity analysis of local flexibilities for voltage regulation in unbalanced LV distribution system," *IEEE Eindhoven PowerTech*, pp. 1–6, 2015.
42. L. Urquiza-Aguiar, C. Tripp-Barba, and M. Aguilar Igartua, "A Geographical Heuristic Routing Protocol for VANETs," *Sensors*, vol. 16, no. 10, pp. 1–28, 2016.
43. K. Ravikumar, and T. Vishvaroobi, "Congestion Control in Vehicular Ad Hoc Networks (VANET) Using Meta-Heuristic Techniques," *International Journal of Computer Science Trends and Technology (IJCST)*, vol. 5, no. 4, pp. 66–72, 2017.
44. R. Saxena, M. Jain, D.P. Sharma, and S. Jaidka, "A review on VANET routing protocols and proposing a parallelized genetic algorithm-based heuristic modification to mobicast routing for real-time message passing," *Journal of Intelligent & Fuzzy Systems*, vol. 36, no. 3, pp. 2387–2398, 2019.
45. S.P. Sasirekha, P. Amudhavalli, and P. Sherubha, "Efficient Route Map-based Ant Colony Optimization for Route Discovery in VANET," *International Conference on Communication and Signal Processing (ICCSP)*, pp. 455–459, 2019.
46. O. Senouci, S. Harous, and Z. Aliouat, "A New Heuristic Clustering Algorithm Based on RSU for Internet of Vehicles," *Arabian Journal for Science and Engineering*, pp. 1–19, 2019.

5

Recent Advancements and Research Challenges in Design and Implementation of Autonomous Vehicles

Mohit Kumar* **and V. M. Manikandan**

Department of Computer Science and Engineering, SRM University-AP, Andhra Pradesh, India

Abstract

Autonomous vehicles (AVs) are one of the most trending technologies these days. Mostly, these vehicles are automated based on machine learning (ML) models and artificial intelligence (AI) approaches. Moreover, many automotive giants have sprung into the market to create a version of autonomous vehicles that any general user can use. Through this chapter, you will come across the history and the experiments which laid down the grounds for the existence of self-driving cars (or autonomous vehicles). You will come across the technologies that help in building up an autonomous vehicle. This module will also illustrate how artificial intelligence, machine learning, neural networks, and the internet of things (IoT) play a significant role in resolving the major challenges associated with self-driving cars. Moving further in the chapter, you will comprehend the several levels of operating modes present in an autonomous vehicle, starting from level 0 to level 5. Later on, we will take you through the safety measures taken care of while building an autonomous vehicle. The book chapter consists of several schematic representations of autonomous cars that will assist you in developing a crisp understanding of the working of autonomous vehicles. The module will draw attention to the peculiarities of autonomous vehicles like intelligent traffic control, adaptive routing, diminishing congestion in areas, traffic signs detector, object detection, speed manipulation, auto-healing, theft control, warning related to failures, etc. Later, the chapter discusses the challenges in launching a self-driving car. You will dive deep into the working principles and the advantages and disadvantages of

**Corresponding author*: mohitkumar_mahto@srmap.edu.in

Romil Rawat, A. Mary Sowjanya, Syed Imran Patel, Varshali Jaiswal, Imran Khan and Allam Balaram (eds.) Autonomous Vehicles Volume 1: Using Machine Intelligence, (83–112) © 2023 Scrivener Publishing LLC

the actual implementation of autonomous vehicles. Security, safety, design, performance, and accuracy are significant challenges associated with any self-driving car. Finally, we will have a brief discussion about the companies that are planning and building autonomous vehicles. The growth and the future scope of autonomous vehicles in different regions of the world are also briefly described along with some essential aspects of the market.

Keywords: Autonomous vehicles, machine learning, artificial intelligence, intelligent traffic control

5.1 Introduction

The era of artificial intelligence (AI) and machine learning (ML) has brought several advancements in science and technology. Autonomous vehicles (AVs) are one such example of the application of ML and AI. Autonomous or self-driving cars are the new generation vehicles that take much of the load out of the driver's hands. It can sense its environment, and based on the environment, it can decide to proceed. Autonomous vehicles can detect the presence of objects, humans, obstacles, etc., with the assistance of computer vision (CV), deep neural networks (DNNs), LiDAR, RADAR, GPS, odometry, etc.

According to the Society of Automotive Engineers (SAE), autonomous vehicles have six classes. Starting from level 0 to level 5 [1]. Level 0 indicates that the driver has the most control over his car, and the vehicle may only turn in a few warnings to the driver while he is driving. Whereas when you move further to level 5, you can experience full automation of the vehicle, where the passengers and the driver just need to sit back in their car and set the destination. In level 5, the car is capable of handling and making necessary decisions to reach the goal.

Autonomous vehicles are designed to take feeds from the sensors and use some regression algorithms to decide its path. Usually, self-driving cars use Bayesian regression, neural networks regression, and decision forest regression to analyze the data provided by the sensors and try to figure out and distinguish the objects that come in their paths [2]. Object detection is one of the significant steps in building an autonomous vehicle. Your autonomous vehicle should be able to distinguish between people, objects, and traffic signs. Based on the traffic signs and signals, the car can abide by the rules and prevent accidents. ISO26262 and ISO/PAS 21448 are the two primary security standards used in maintaining the safety of electrical and electronic components present in an autonomous vehicle [3].

5.1.1 History and Motivation

Autonomous vehicles as an idea had an existence way back in the 1920s. In the 1980s, the first self-driving car came into existence with the University of Carnegie Mellon University's ALV project. Later on, in 1984, Mercedes Benz and the University of Munich also came up with an autonomous vehicle. In August 2013, Daimler research and development collaborated with Karlsruhe Institute of Technology/FZI [4] to create a Mercedes-Benz S-class car that drove entirely autonomously for roughly 100 kilometers from Mannheim to Pforzheim, Germany, following the famous Bertha Benz Memorial Route [5]. In current scenarios, Tesla is emerging as one of the companies launching self-driving cars with autopilot modes. In the 100 years from the 1920s to the 2020s, several challenges occurred, some of which still exist in the field. Conquering these challenges is one of the biggest obstacles in laying down the ground for autonomous vehicles. Even companies like Google, Microsoft, Tesla, IBM, etc., are advancing the existing DNN models that can help overcome the challenges existing in autonomous vehicles. Autonomous vehicles bring a wide variety of research challenges, and while solving them, you can come up with a product that can be used worldwide.

5.1.2 Present Scenario and Need for Autonomous Vehicles

Nowadays, if you notice the reports of WHO, then you will find that around 1.3 million people a year die due to road accidents [6]. If you closely observe the age group involved in road accidents, you may get shocked as most are between 5 and 29 years old. There are scenarios where people are found driving drunk, drugged, or inexperienced in tackling upcoming obstacles. Even trained drivers sometimes make a mistake which leads to injuries and sometimes death. Driving a vehicle on the road is not just a concern for our own lives or safety, but it's also a concern of others' safety too. On roads, there are other vehicles, pedestrians, poles, etc. A slight mistake may lead to a safety concern for both us and the others on the road.

The idea of autonomous vehicles brings simplicity and removes a lot of the burden from a driver's hands. Vehicle driving is a skill that needs time and effort to master. If we can train a human to drive, we shall try the same with a machine [9]. The thought of training machines to operate themselves brought the autonomous vehicle into existence. It may not completely resolve the problem of road accidents, but these machines can indeed reduce the rate of road accidents if appropriately developed.

5.1.3 Features of Autonomous Vehicles

Self-driving cars or autonomous vehicles put a lot of features in the hands of the driver. Driver have several modes to operate their vehicle. They may decide which mode they can use to reach their destination. Moreover, the autonomous system handles traffic quite intelligently. It gets to its destination efficiently as the autonomous vehicles are loaded with maps that adapt to the current traffic situations and self-determine the best path to reach the goal. Self-driving cars reduce traffic congestion. Autonomous vehicles collect data from each vehicle's working parts and analyze if any failure or damage is there. It generates warnings to the users if some loss or damage occurs. Autonomous cars are trained to strictly follow the traffic rules and signs which ensure road safety. The features of autonomous vehicles are stated in Figure 5.1.

5.1.4 Challenges Faced by Autonomous Vehicles

Safety is a vital concern for any autonomous vehicle. Without ensuring safety, autonomous vehicles cannot come onto the market. Drivers are usually worried about how the speed and the torque on the steering wheel can be adjusted depending on the type of roads encountered. Nowadays, there are trust issues on the ML models, whether the models would work

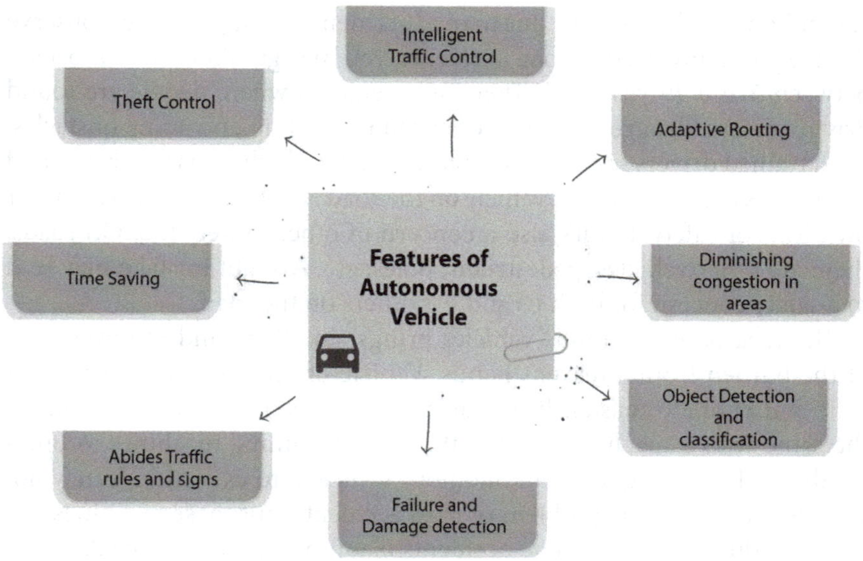

Figure 5.1 Features of autonomous vehicles.

efficiently or not. Coming to the fact that these kinds of vehicles are primarily dependent on sensors, like cameras, Radar, Lidar [7], now the makers need to ensure that there is no delay between transmitting the collected data from sensors and generating the response from the model. If the delay is significant, it may lead to delayed reactions, leading to severe accidents. Imagine a scenario where you are riding in an autonomous vehicle that is driving you to your destination. Suddenly a pedestrian appears as if from nowhere in front of your car. In such a scenario, if the sensor delays in sending the data or the model delays in predicting the action of braking, it will result in a loss of life.

5.2 Modules/Major Components of Autonomous Vehicles

The major modules of autonomous vehicles are discussed in this section.

5.2.1 Levels of Autonomous Vehicles

Autonomous vehicles can be further classified based on their levels of automation. In any general autonomous vehicle, there are six levels of automation [1]. These levels are like the modes of operation for an autonomous vehicle. Let's look at each of these levels and understand the facilities each group offers.

- Level 0 (No Automation): There is no automation at this level. The master control of the vehicle is in the hands of the driver. The driver needs to decide the route they wish to take and drive that route by themselves.
- Level 1 (Driver Assistance): At this level, there are features like adaptive routing, traffic control, location tracking, and warnings for drivers. Using such facilities, the driver can leave a few things like deciding the best route to reach the destination to the vehicle itself. The car can safeguard the driver by assisting and providing warnings whenever required.
- Level 2 (Partial Automation): The vehicle can accelerate or decelerate itself based on the road conditions. Along with this, the car has a hold on the steering wheel and can monitor it to maintain the vehicle's path.

- Level 3 (Conditional Driving Automation): At this level, the vehicle is capable enough to capture information from its surroundings and then make a decision. It can sense pedestrians and objects present on its path. The car comes up with several sensors that help in tracking the surroundings.
- Level 4 (High Automation): At this level, there is almost full automation from the vehicle but there is some input from the driver. You can customize the limits and even override the decisions taken by the car on the way. Suppose, if the vehicle says to go at 60km/hr speed in a particular place, you can override that decision and speed up to 70km/hr.
- Level 5 (Full Automation): This is the topmost level of automation that any user can expect from an automated vehicle. At this level, you can expect the car to be 0% dependent on the driver. You just need to sit back in your car and fix the destination, and the rest will be decided by the vehicle itself. Here, drivers cannot override vehicle decisions, which is why it is called full automation. Figure 5.2 shows about the representation of levels of automation present in autonomous vehicles.

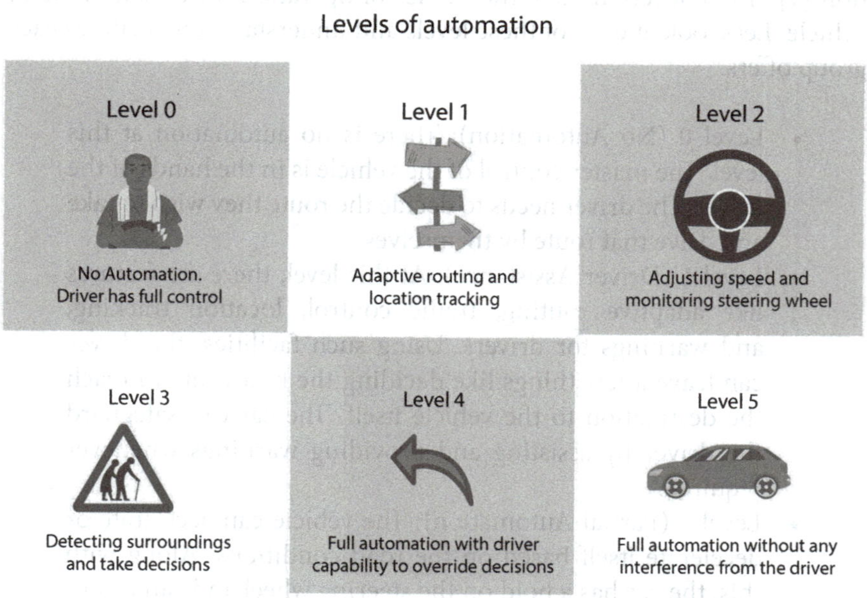

Figure 5.2 Levels of automation present in autonomous vehicles.

5.2.2 Functional Components of An Autonomous Vehicle

Autonomous vehicles are possible due to the presence of some advanced components and sensors. The technical advancements in electrical and electronic components led to the generation of some advanced sensors and devices that assist in designing a self-driving car. The presence of such devices ensures driver safety and accident avoidance. The automotive expert Robson Forensic has done detailed research on the variety of functional components that assist in designing and providing performance to autonomous vehicles [8].

For any general autonomous vehicle, the following are some of the crucial aspects which it should try to provide:

- Adjusting speed and acceleration of the vehicle
- Monitoring the direction of the vehicle by using the steering wheel
- Applying brakes when the vehicle senses the destination has been reached or some unforeseen situation has occurred
- Should detect and distinguish objects present on the road
- Tracking the best route and reducing the traffic

The above functionalities, in reality, are possible due to the presence of numerous sensors and devices. Figure 5.3 below shows a brief summary of the sensors and devices present in an autonomous vehicle.

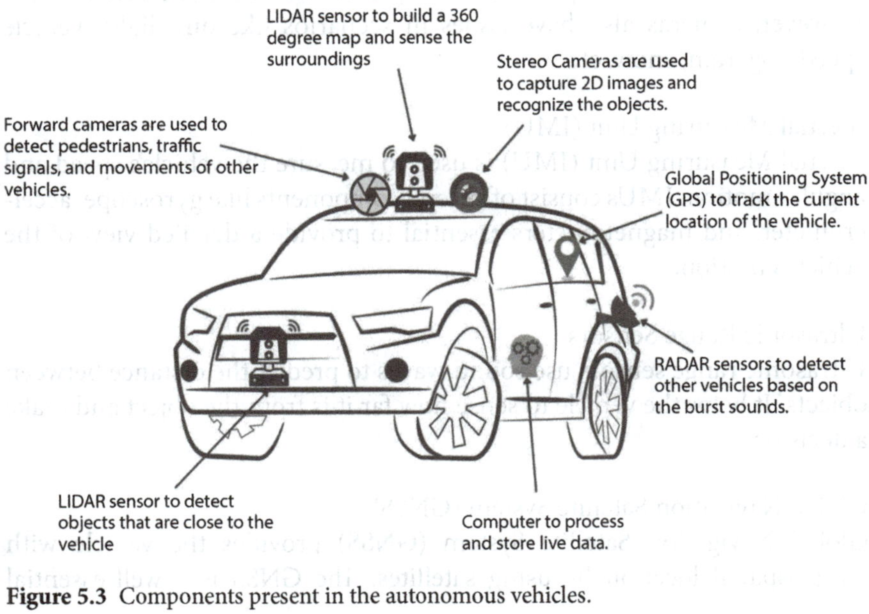

Figure 5.3 Components present in the autonomous vehicles.

Now, let's focus on each device in detail and understand their working and use case.

Radar

The radar sensor uses sound signals to detect other vehicles and objects present on the road. It emits a sound signal and waits for some time for the signal to revert. When the sensor receives back the sound, it senses the distance at which the other vehicle is. Radar works in conditions like dense fog, rain, and even snow. Using Radar, we can measure the relative speeds of the cars.

Lidar

Lidar (or Light Detection and Ranging) is a well-known sensor similar to that of radar. However, in the case of Lidar, lasers are used to detect objects instead of sound signals. Lidars are comparatively better in terms of range and provide clarity of the things present nearby the vehicle. It is capable of sensing a 3-dimensional map of the surroundings. Lidars are expensive and also face issues in scenarios like dense fog, rain, and even snow.

Cameras

Radar and Lidar sense objects whereas cameras help in capturing the surroundings. While the vehicle is on the road, its task is not just to sense objects but also to distinguish them. Cameras primarily capture 2D images in frames that are later provided as input to the ML and DNN models. However, cameras also have issues in scenarios like dim light, vehicle speed, fog, rain, snow, etc.

Inertial Measuring Unit (IMU)

Inertial Measuring Unit (IMU) is used to measure the vehicle's speed and angular motion. IMUs consist of several components like gyroscope, accelerometer, and magnetometers essential to provide a detailed view of the vehicle's motion.

Ultrasonic Range Sensors

Ultrasonic range sensors use sound waves to predict the distance between objects. It helps the vehicle to sense how far it is from the object and make a decision.

Global Navigation Satellite System (GNSS)

Global Navigation Satellite System (GNSS) provides the vehicle with its geospatial location by using satellites. The GNSS is a well-essential component of autonomous cars as without knowing the vehicle's current location, the car can't predict its route to reach the desired destination. GNSS is capable of working in any kind of weather condition and provides a highly accurate location.

5.2.3 Traffic Control System of Autonomous Vehicles

Autonomous vehicles have a component to monitor the traffic and safeguard the car from entering congested areas. Congested areas, as we know, have several objects, cars, and pedestrians, which creates difficulties for the sensors and devices to locate them and make judgments. Entering a congested route may also lead to delay in reaching the destination. Even for a regular driver, the situation is very hectic. When the road is too clogged, even a slight displacement must be taken carefully; even a sudden rise in speed may lead to collisions.

To avoid traffic congestion, a traffic control approach based on predicted findings is used. You can find two separate traffic volume and speed forecasting algorithms to determine the amount of congestion in a sector. The prediction model for traffic volume is used to see if traffic volume variations were causing congestion, and the predictive model for speed is used to figure out how bad it is. The suggested system may forecast and mitigate traffic congestion by learning AVs' geographically and temporally congested traffic patterns. As a result, adequate AVs transport capacity is provided, and steady production is achieved under dynamic situations [17].

5.2.4 Safety Features Followed by Autonomous Vehicles

Safety is one of the crucial concerns for autonomous vehicles, and their manufacturers always try to provide all possible safety features. In the current scenario, the following safety features may be a part of several autonomous vehicles.

1. Adaptive Cruise Control (ACC)
   Adaptive cruise control provides speed monitoring services for autonomous vehicles. It detects both the front and rear of the car and tries to adjust a perfect speed to go on in that particular route. By keeping the distance between the vehicles, ACC also prevents collisions and leads to a safe journey. If the road seems empty with very few cars, it speeds up the

vehicle or slows down the car. So, depending on the situation, the ACC adjusts the speed parameters.

2. Autonomous Emergency Braking (AEB)
   An autonomous emergency braking system scans the entire route beforehand and locates the positions where it may need to apply brakes. It also has advanced methods to detect what is near the vehicle and make sure there is no collision, and it automatically applies the brakes.
3. Electronic Stability Control (ESC)
   Electronic stability control is a component of the autonomous vehicle that provides stability to the vehicle while driving under critical situations. If the road is imperfect, it tries to determine the speed at which it will experience fewer jolts and ensures security on such roads. While applying brakes, it looks into the condition of the road and makes sure that the vehicle does not lose its stability on applying the brakes.
4. Reverse Park Assist (RPA)
   Reverse park assist is an ubiquitous component that nowadays even the standard vehicles carry. RPA, as the name suggests, is a tool that assists you while driving backward or while parking your vehicle. You can see your front view and easily navigate, but it's not easy to just look at the rear side thoroughly. So, modern cars and even autonomous vehicles have come up with reverse park assist that uses cameras and algorithms to detect the vehicles' objects. It also generates warnings if your car is very close to something.
5. Lane Keeping Assist (LKA)
   Driving in the correct lane is quite essential for vehicles to avoid accidents. Lane-keeping assist helps vehicles in maintaining the correct lane throughout the journey. It tries to adjust the torque and the angle at which the steering wheel should be rotated to be in the right lane. Detecting lanes is quite an essential step for the vehicle to adjust the vehicle's direction.
6. Rear Cross-Traffic Assist
   The rear cross-traffic assist is quite similar to the reverse park assist model. However, this model is more advanced as it detects traffic at the rear side of the vehicle and assists the driver whenever he needs to reverse. While driving your car

on the road, you may encounter traffic, and sometimes, to clear that traffic, you need to go backward. Now the problem arises when there is also a bit of traffic, so this system helps you detect the rear traffic and suggest the path to take the car backward.

7. Vehicle-to-Vehicle Communication
   Vehicle-to-vehicle communication is quite an essential component of modern autonomous vehicles. Now the vehicles themselves can communicate with each other to find out the correct speed, position, and direction they need to follow. While on the road, the vehicles can use the information from other vehicles about the upcoming traffic and road conditions. Imagine a scenario where you are going up the road and some vehicle is coming down the road. In such scenarios, your vehicle can communicate with the oncoming vehicle about the traffic ahead and its speed ahead of the route.
8. Vehicle Guidance System
   A vehicle guidance system is like a monitoring system for every part of the vehicle. It is like a human driver who knows everything about their car and the destination they need to reach, and the route they need to follow. The vehicle guidance system looks into tasks like monitoring the sensors, collecting data, looking after the engine status, planning the best route for the journey, and providing warnings to the users (if required).
9. Traffic Jam Assist
   Nowadays, autonomous cars are powered up with traffic jam assist tools that predict the traffic and save you from getting into traffic jams. However, if you enter such a zone by any chance, these systems help you by rerouting your journey with a route free from traffic. It also works coupled with your adaptive cruise control to adjust the speed based on the traffic conditions.
10. Blindspot Detection
    For a driver, it's not possible to have a 360-degree view of the surroundings while driving the car. The driver can't perfectly see what's happening at the rear or side of the vehicle. It has rearview mirrors which give information about the objects appearing from the rear. However, these mirrors are not sufficient to have a complete view of the rear. It's not just the

rear that may create trouble. Even at the sides of the vehicle (left and right), also up and down, there may be some objects which a driver may miss. Such areas are known as blind-spots which the driver may tend to not know about or overlook. In such cases, blindspot detection is one of the saviors. Blindspot detection tries to keep an eye on the 360-degree view of the car's surroundings and captures all the necessary information for driving the vehicle. Just imagine you are trying to reverse your car, and suddenly you can't see a ditch at the rear side of your vehicle, due to which your vehicle may fall into that ditch, and this may result in vehicle collapse or some machinery failures.

5.3 Testing and Analysis of An Autonomous Vehicle in a Virtual Prototyping Environment

You cannot always test your models and the car directly on the roads. It is risky and also needs a hefty amount to test all your prototypes. Imagine a scenario where you came up with a new algorithm and wanted to try how it would work. One thing you can do is spend a considerable amount and build a car following your set of algorithms and then test it on the road. The vehicle may work perfectly, but mostly, that is not the case. You may even see your algorithm failing, or you can notice some loopholes in it. Now, since you have already implemented the algorithm, you can no longer change the existing system. It's a loss of money, effort, and time. So, to prevent such losses, you can use some simulation models which create a virtual environment resembling the exact road scenarios. You can test your AI/ML models and check the sensors and other devices you feel your car carries using such simulation models.

As discussed before, machine learning components like neural networks, for which formal qualities are difficult to demonstrate, are one of the key obstacles in testing autonomous driving systems. Sim-ATAV is a simulation-based adversarial testing platform that enables evaluating Cyber-Physical Systems, such as automated falsification driven by formal requirements and search-based test case creation [18]. It is an add-on for the Matlab toolbox S-TaLiRo that simulates cars using the open-source Webots robotics simulator. You can see Figure 5.4 to understand more about the simulation model and its working.

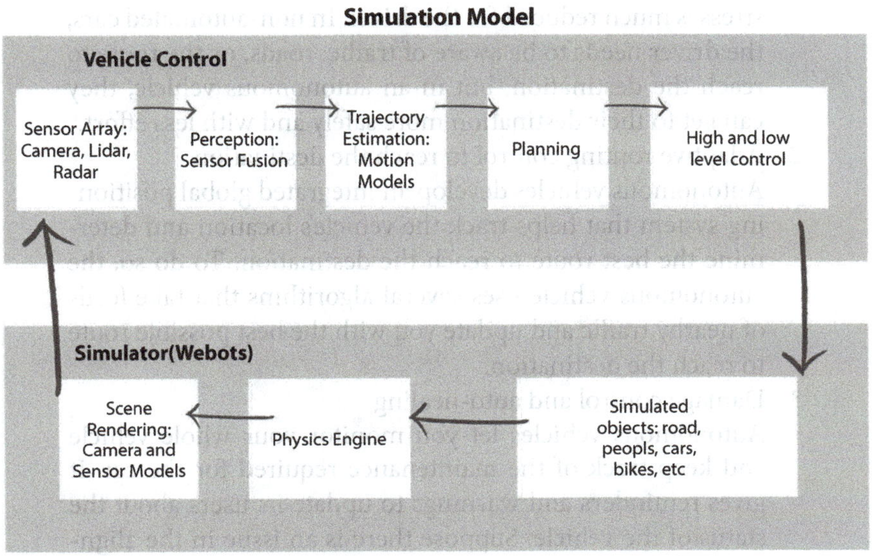

Figure 5.4 Working of Sim-ATAV simulation model.

The perception system processes data from three sensor systems: CCD camera pictures, lidar, and radar. The framework runs on a typical desktop PC and makes use of publicly accessible and low-cost tools. Later, Sim-ATAV may be used to construct classic and sophisticated automated testing methodologies that were previously impossible to execute with existing frameworks.

5.4 Application Areas of Autonomous Vehicles

Autonomous vehicles are not just meant to carry people from one place to another. Even a standard car or public transport can do that. Autonomous vehicles themselves bring several new technological advancements and driver assistance systems with many real-life applications. Some of them are listed below.

1. Removes the load from the driver's hand
   Autonomous vehicles are driverless and require minimal effort to drive out to the destination. Driving is a skill and requires practice and effort to become proficient. People spend lots of money and time in attaining such skills. However, with the coming of autonomous vehicles, the

stress is much reduced for the driver. In non-automated cars, the driver needs to be aware of traffic, roads, or the route to reach the destination, but in an autonomous vehicle, they can get to their destination more safely and with less effort.

2. Adaptive routing control to reach the destination
   Autonomous vehicles develop an integrated global positioning system that helps track the vehicle's location and determine the best route to reach the destination. To do so, the autonomous vehicle uses several algorithms that take feeds of nearby traffic and update you with the best possible route to reach the destination.
3. Damage control and auto-healing
   Autonomous vehicles let you monitor your whole vehicle and keep track of the maintenance required for the car. It gives reminders and warnings to update its users about the status of the vehicle. Suppose there is an issue in the alignment of the wheels, then the car can detect it and update you to take the necessary action.
4. Intelligent Traffic control
   Traffic is one of the most significant issues which anyone faces while traveling. Sometimes the traffic may result in delays in reaching the destination. Safeguarding the vehicle from crowded areas would not only save time but would smooth the journey. Autonomous vehicles try to use some specialized routing algorithms that assist them in deciding the route and speed with which they should proceed. As a result of this, the autonomous vehicle assists in reducing congestion in areas.
5. Fewer road accidents
   Humans may break the rules, but machines are not trained to break the rules; they follow the traffic rules and the safety guidelines strictly. Since the decisions are made by the car itself with less or even zero human interference, fewer rules are broken. Most of the accidents take place due to breaking the rules or following some unethical way of driving. On the other hand, autonomous vehicles always stick to the rules and follow the general pattern of driving, which makes them more secure from encountering road accidents. However, it is still a machine, and it may even lead to an accident in some unexpected situations.

5.5 Artificial Intelligence (AI) Approaches for Autonomous Vehicles

The major use of AI in autonomous vehicle design is discussed in this section.

5.5.1 Pedestrian Detection Algorithm (PDA)

Pedestrian detection is one of the essential sections of an autonomous vehicle. Although several sensors collect feeds from the surrounding environment and sense the object, detecting pedestrians is crucial. Roads are not just meant for vehicles. People and other vehicles are also a part of the same road on which your vehicle is traveling. The pedestrian detection algorithm is one of the applications of AI and ML, which is widely used in surveillance systems. Nowadays, this algorithm plays a significant role in advancing the path for autonomous vehicles.

The ultimate goal of the pedestrian detection algorithm is to collect input images from the cameras available on the body of the car, then process it and figure out the pedestrians in that image. You can think about this step more by understanding a real scenario. Just suppose that you are on your way to some destination using your autonomous vehicle. On the way, you may encounter several pedestrians who are walking along the road. You can easily recognize that some pedestrians are walking in front of you, but the vehicle can't directly recognize pedestrians or ordinary objects. In that scenario, you as a person may try to lower the speed of your car or even apply brakes if you find pedestrians close to your vehicle. For autonomous vehicles, this task is done by the pedestrian detection system, which detects pedestrians and then decides whether to lower speed, apply brakes, or do something else. Figure 5.5 represents for the autonomy for analysing autonomous vehicle detecting pedestrians on their way.

Figure 5.5 Autonomous vehicle detecting pedestrians on their way.

The pedestrian detection system is implemented by using both traditional as well as DNN approaches [10]. Let's examine each of them separately and understand the algorithms provided by each of them.

Traditional Approach

In the case of the traditional approach, an image is provided as input. Using this input image, the model tries to extract some features. A human body comprises the head, hands, legs, neck, etc. These body parts are the features which the traditional algorithm will try to figure out. The features are then used to train the classifiers to differentiate between the pedestrian and other objects present in the road. The traditional approach is described using the following flowchart (Figure 5.6).

Modern approach using HOG features and SVM classifier

As time progresses, the industry will come up with better approaches to solve a current issue. The issue of detecting pedestrians can be solved using a traditional system. However, that method is not that reliable and fast. Nevertheless, a brilliant and reliable way of detecting pedestrians is HOG (Histogram of oriented gradients), which is one of the most commonly used features for detecting pedestrians. Slopes accurately reflect

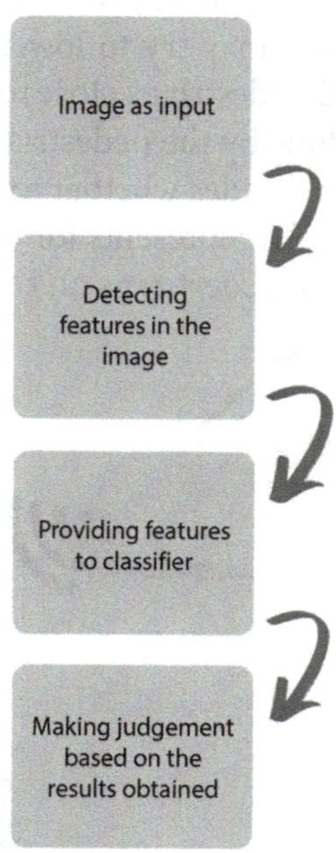

Figure 5.6 Flowchart of the traditional approach of detecting pedestrians.

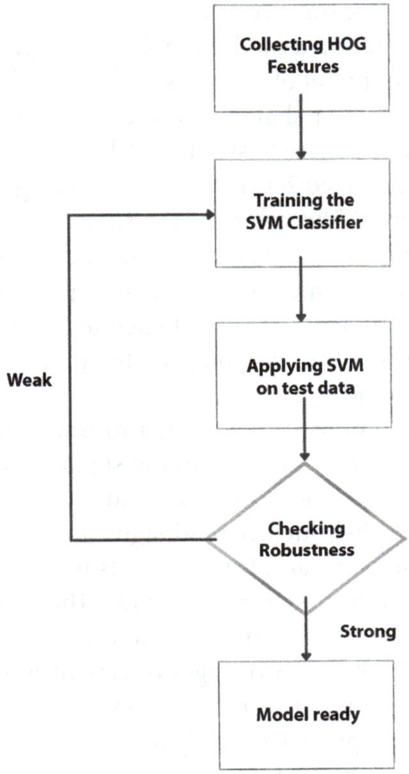

Figure 5.7 Flowchart of the modern approach of detecting pedestrians.

the representation and structure of local objects (i.e., edges) in pictures, and gradient direction histograms of local regions in images can be used to generate features [10]. SVM (Support vector machine) is a classifier algorithm that takes the features as input and classifies them into various categories of objects. Figure 5.7 is a schematic diagram to represent the algorithm.

5.5.2 Road Signs and Traffic Signal Detection

Road signs and traffic signals are an essential part of the experience of driving a vehicle on the road. There are some rules and guidelines set down by the government or the road safety department whose practices every driver needs to follow. Coming up with autonomous driverless cars requires some other ways to follow such rules. The usual rules put down on paper can be coded directly, and the vehicle will follow them. The main issue arises when we have some restrictions based on the road or the traffic. For example, when you find a road symbol that designates not to blow any horns, the vehicle should not blow horns in those areas. If the traffic signal indicates a red light, the car shall stop there and wait for the light to turn green. Road signs must adhere to strict guidelines when they are designed, produced, and erected. Fixed 2-D forms such as triangles, circles, octagons, and rectangles are used. The colors of the signs are selected to stand out from the surrounding landscape, making them easily identifiable by cars. One color is used for the information on the sign, while another is used for the remainder of the sign. The use of fixed text typefaces and character heights distinguishes the traffic signs [11]. Figure 5.8 represents distribution of some general road signs.

Drivers and autonomous vehicles need to understand and should be able to recognize the road signs and traffic signals present in such areas. The DNNs, image recognition system, see and image segmentation are the core components that help detect road signs and classify traffic signals. The detection of road signs and traffic signals is not a new topic and has a greater area of research from earlier periods. The method to detect road signs is the same as applying computer vision to detect some objects and then classify them. There are two stages of classification of those images; first is the detection, and then is recognition [11]. Detection phase deals with capturing and the preprocessing parts. It detects the essential areas and performs image segmentation to get functional properties like color

Figure 5.8 Some general road signs.

and shape, which are indispensable for any road sign. The output of the detection phase then goes to the recognition phase, where a pre-built model tries to compare the properties like color and shape and then maps it with a highly probable road sign from the dataset. An advanced way of classifying the road sign is by pictogram analysis, which focuses on the text inside the sign and tries to map it with a perfect class of road signs. Figure 5.9 describes the model and the algorithm in a clear and precise way.

Traffic signals have three different lights, namely red, yellow and green. To detect such colors and classify them as red, yellow, and green, RGB cameras capable of capturing colored images are needed. Color-based detection is not easy for any model. However, the approach considers other facts, like ambient lights, which may affect the signal detections. Overlapping of colors and the presence of black corners are problems that can occur while detecting traffic signals. Color space conversion, or translating the RGB image into a form that facilitates the detection process, is vital for a color-based detection system. It entails separating the color information from the brightness information by transforming the RGB color space into another color space that provides strong detection skills depending on the color cue.

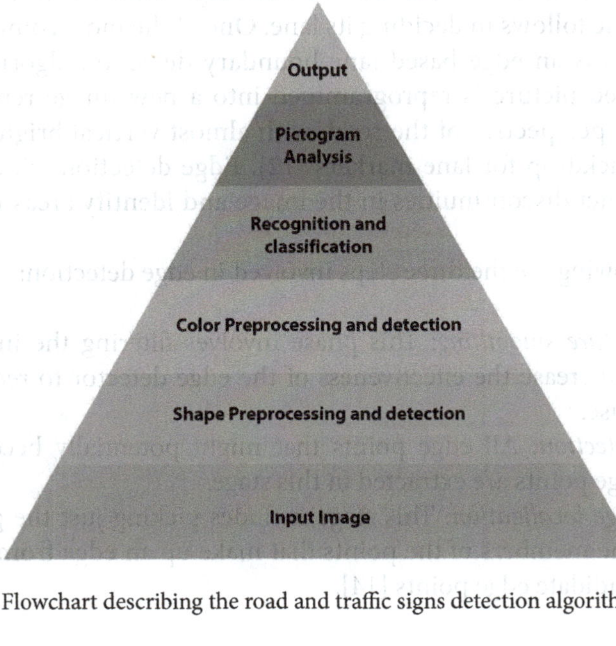

Figure 5.9 Flowchart describing the road and traffic signs detection algorithm.

5.5.3 Lane Detection System

Roads are of various types and have their characteristics, making them separate from other kinds of roads. You must have seen roads with white broken lines, solid yellow lines, continuous white lines, etc. These kinds of colors and patterns are made on the streets to distinguish them. Drivers need to observe the type of road and the drive. For example, let's consider the double continuous yellow solid lines on the road. Such a pattern designates that you cannot overtake, change lanes or take a U-turn on such roads. Similarly, for other patterns, we have a specific rule. Every vehicle that travels such roads abides by the laws and regulations decided for that particular road. Suppose you are going through a street where changing lanes is not allowed, but for any reason, you changed the lane, and the other cars, unaware, didn't change their lane. In such cases, it may lead to accidents or the collapse of two vehicles. So it becomes essential to observe the lane and road patterns, to make sure you are not going on the wrong track.

Autonomous vehicles themselves are empowered with advanced lane detection systems. Such systems help in recognizing road patterns and then deciding the lane on which the car should follow. Failing to do so may result in choosing the wrong route and may result in some accidents. To ensure higher safeties and proper lane detection, several algorithms are proposed. Here you will see one such algorithm which an autonomous vehicle follows in deciding its lane. One of the most common algorithms used is an edge-based lane boundary detection algorithm [13]. The captured picture is reprogrammed into a new image representing a bird's eye perspective of the road, with almost vertical bright lines on a darker backdrop for lane markers [12]. Edge detection tries to figure sharp, distinct discontinuities in the image and identify areas with solid intensity.

The following are the three steps involved in edge detection:

1. *Picture smoothing*: This phase involves filtering the image to increase the effectiveness of the edge detector to reduce noise.
2. *Detection*: All edge points that might potentially become edge points are extracted in this stage.
3. *Edge localization*: This stage includes picking just the genuine members of the points that make up an edge from the candidate edge points [14].

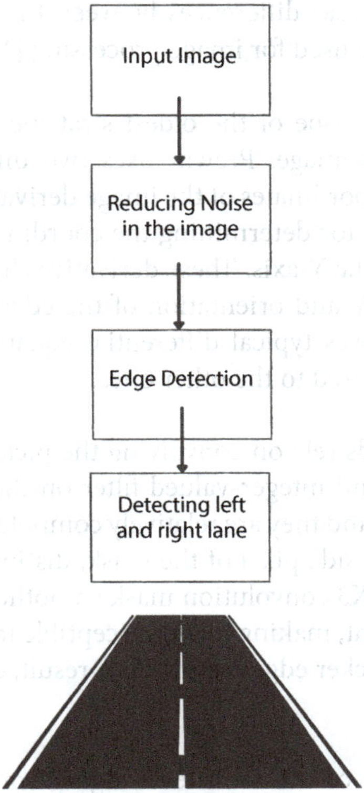

Figure 5.10 Lane detection algorithm involving noise reduction and edge detection.

This may be clearly understood by observing the following flowchart (Figure 5.10).

Noise reduction and edge detection are two crucial steps for the algorithm. If the image has too much noise, then it may not ideally determine the lane. Along with these two, one more significant step is to detect the vehicle's relative position along the road. The algorithm should keep track of its relative positions and turn the steering wheel to ensure that the vehicle falls within the prescribed lane. Some of the standard techniques which make up the lane detection system are as follows:

1. Roberts
   One of the oldest techniques of implementing lane detection is Roberts. In these techniques, the difference between the set of mutually perpendicular directions is considered a

gradient, whereas differences between the diagonally adjacent pixels are used for image processing [14].

2. Prewitt

   Prewitt is also one of the oldest strategies to detect edges from a given image. Prewitt uses two different masks to identify the coordinates of the image derivative at the X-axis and the other for determining the coordinates of the image derivative at the Y-axis. These derivatives help in estimating the magnitude and orientation of the edges. However, this method involves typical differential equations and is a bit tedious compared to the other ones.
3. Sobel

   These methods rely on convolving the picture with a small, extractable, and integer-valued filter on the horizontal and vertical axes, and they are relatively computationally efficient [15]. At the inside pixel of the mask, distinctions are determined. The 3X3 convolution mask smoothes the image by a certain amount, making it less susceptible to disturbance. In any event, thicker edges result. As a result, edge detection is inadequate.
4. Canny

   Canny is one of the multilevel algorithms that is a well-known application of computer vision to detect edges from an image. If the gradient amount is more than that of pixels on both sides in the direction of intensity change, this approach classifies it as an edge. Smoothing the picture to remove noise is part of the procedure. The picture gradient is then found, and spatial derivatives are used to indicate the region of interest. These areas are then pinpointed.

5.6 Challenges to Design Autonomous Vehicles

Autonomous vehicles are still not so standard in a few countries. In many places, autonomous cars are operating on the roads with some of the other restrictions. An autonomous vehicle includes several sensors, modern technologies, machinery, IoT devices, AI/ML/DNN models, etc. The longer the list, the more the challenges in handling them. Let's briefly look into some of the difficulties which usually occur while designing or implementing an autonomous vehicle.

1. Synchronization
   Synchronization is quite a critical step for any autonomous vehicle. If the sensors and the vehicle's machinery are not properly synchronized, it may be adverse. Just imagine an autonomous vehicle going on its way to its destination. Suddenly, as if from nowhere, a child steps in front of the car. If the machinery and the sensors are not synchronized, it may lead to accidents and other adverse situations.
2. Speed
   Speed is essential for any vehicle. Depending on the traffic and road conditions, the rate of the car should automatically adjust. Speed is one of the most critical concerns for designing any autonomous vehicle. When you pass through a crowded place with no road sign determining the speed, the car should itself predict the rate with which it should proceed further.
3. Diverse weather conditions
   Weather is one such condition that is never the same, and you cannot expect every day to be sunny and clear. At some places, you may even find rain, snow, or fog. The problem arises when the vehicle goes through changing weather conditions. The sensors and the cameras should be capable enough to adapt to the changing weather conditions. Due to changing weather conditions, the sensors and cameras may fail to detect the nearby objects, and the vehicle may collapse. One such scenario may happen when there is dense fog present. Suppose, due to thick fog, your cameras couldn't capture the surrounding adequately or detect the road signs appropriately. In such scenarios, the vehicle may make some wrong decisions and may lead to accidents.
4. Variety of objects present
   Roads are not just for cars. There are also pedestrians, traffic signals, road signs, other vehicles, electric poles, etc. Sometimes it is too hard to train the car for all types of objects that it may encounter. You never know the objects for which the vehicle is prepared, only that things may appear in front of the vehicle. When the car moves on, it may be ready to detect traffic signals and other cars. Suddenly, say, a baby stroller came near your vehicle. If the vehicle doesn't

detect it as a meaningful object, it may hit it. So, the variety of possible obstacles creates a lot of issues for a car to figure out whether to stop or ignore the object on the road. Sometimes due to road repairs or other repair works, there are road change signs or other signs. If the vehicle does not understand those objects, it may go onto a broken road or one that is under repair.

5. Diversity of pedestrians
   People belonging to various communities, cultures, traditions, etc., may have different appearances. Now the problem is how to distinguish them. There are several figures, shapes, and types of clothing in which a pedestrian may appear in front of the vehicle. The algorithm, even though robust, may not be 100% accurate in detecting the pedestrian appropriately because of the changing characteristics.
6. Different illumination conditions
   Whether day or night, the sensors and the cameras should work efficiently and guide the vehicle accordingly. Illumination is significant for any camera to capture pictures precisely. However, it's not always guaranteed that the car will receive an ample amount of light to capture the images. It may be evening or night where there are very few lights. By any chance, if you are driving in an area with very few street lights, then it becomes difficult for the cameras and sensors to figure out the surroundings around your vehicle.
7. Radar and Lidar interference
   Most variants of autonomous vehicles are equipped with Radar and Lidar. There is no problem when only a few autonomous cars are on the roads. However, in the future, there will be many autonomous vehicles on the streets. The real problem arises when, say, a car sends a signal from the Radar and waits for it to reflect. Meanwhile, some other vehicles have also sent a signal. Here, it's difficult for vehicles to distinguish between the received signals. It may either be a car's own signal or the signal generated by other cars.
8. Traffic conditions
   Densely populated places and crowded areas are a significant threat for autonomous vehicles, and they need to be very

sharp in figuring out their paths. When traffic conditions are not suitable for driving, the car starts facing difficulties. Imagine a scenario where suddenly due to some situation a lot of vehicles got stuck at a place. Now the authorities need to come and take over the traffic conditions. In such an instance, the authorities make decisions that are new for autonomous vehicles to understand. The algorithm may go to the right but suppose the officers instruct the car to reverse. Then the autonomous vehicle will get confused in making a correct judgment.

9. Road signs may occur in different conditions
   Road signs direct the vehicle about the basic rules that need to be followed on that road. Examples include speed signs and crosswalks signs. Speed signs specify the speed range you shouldn't exceed. Another example is a crosswalk sign, which shows a crosswalk where you need to slow down your vehicle. Road safety departments and governments take initiatives in placing such road signs wherever required. It's always hard to keep track of the way the road signs are on different roads. There are billions of roads across the globe, and keeping an eye on all the road signs is pretty tricky. Sometimes, these road signs get distorted or dismantled due to animals, birds, or even adverse weather conditions. Sometimes people themselves cover the road signs with some posters or advertisements which create difficulties. In such scenarios, the vehicle automatically wouldn't know what speed it should drive or if any crosswalk is there or if it is near a railway track, and may continue at the same speed.
10. Issues with the AI/ML models
   GIGO (Garbage in Garbage out) is a very well-known acronym. The same applies to the AI/ML or the DNN models used for building up the autonomous vehicle. If the input provided to these models is not appropriate, the judgment made obviously would not be reasonable in any terms. While driving at high speed or in a cloudy or foggy environment, it is pretty difficult for the cameras to capture clear pictures. The same blurred or imperfect images will then be given as an input to the model, and the model may predict or classify something based on those pictures. One solution can be keeping more high-tech cameras and capturing each

frame perfectly, even at high speed. However, such cameras would increase the budget for designing the vehicles and make them less affordable.

11. Theft Control

    As you know, autonomous vehicles come up with quite high-tech and costly sensors. Some of its sensors and cameras are continually exposed to outside people. Burglars or thieves can steal such sensors and cameras, which may be revealed with the car. It's pretty challenging to keep track of such items and sometimes very easy for the burglars to succeed. The new generation of autonomous vehicles would surely try to protect the delicate and costlier parts of the car, like sensors and cameras.

12. Trust and Safety issues

    Trust and safety are yet more crucial concerns while developing such vehicles. It is hard to convince people of vehicle safety. A vehicle is a machine and may be subject to accidents or some adverse situations. Nobody can guarantee to eliminate that. While on the road, some sensors, cameras, or other machinery may get damaged and lead to an imperfect journey. However, machinery may also get damaged in a standard vehicle, but making people believe that is truly a difficult task. Recently, if you focus on the news articles, you may find that even level 2 autonomous vehicles have met with an accident. All this news and people's lack of trust creates a big obstacle for manufacturers who wish to lay the ground for autonomous vehicles.

You came across several challenges standing in the way of autonomous vehicles. However, that does not mean autonomous vehicles are just thoughts; they are a reality that will someday find its way into the markets. Many companies and manufacturers are currently working on solving the above challenges and are coming up with some brilliant solutions. Some of the manufacturers who are involved in the development of autonomous vehicles are described in Table 5.1 [16]:

Table 5.1 Companies manufacturing and researching autonomous vehicles.

Company name	**Related works**
Tesla	Tesla has featured autopilot and other assistance features since 2015.
Pony.ai	Pony.ai provides several modern algorithms to solve AI/Ml-related challenges.
Waymo	Waymo is a part of Google and is trying to bring a lot of advances in the field of autonomous vehicles.
Apple	Apple is well known for its Project Titan, which is the electric car powered by Apple. Apple formally revealed in 2016 that it is also working in the field of autonomous vehicles.
Kia-Hyundai	Hyundai and Kia Motors are trying to come up with fabulous autonomous technologies by the year 2030. They are heavily investing in their plans and executing them.
Ford	Ford seems to be a great competitor in the field of autonomous vehicles. Ford is trying to come up with several variants of autonomous vehicles in just a few years.
Audi	Audi, with NVIDIAs AI technology, is coming up with brilliant autonomous vehicle ideas. Audi is already known for providing luxurious cars with several advanced features and will be a great competitor for other manufacturers.
Huawei	Huawei seems to be a supportive company in the market that is trying to collaborate with Audi and promote open-source research work in autonomous vehicles.

5.7 Conclusion

Autonomous vehicles are the future of driving and transportation. Looking forward to the new era, you can surely see several new companies and manufacturers jumping into developing such vehicles. Through this chapter, you came across most of the crucial aspects of an autonomous vehicle. Autonomous vehicles, as you saw, are one of the most prominent applications of the field of AI and ML. You came to know several components, including the sensors and the devices that make autonomous vehicles a reality. Researchers and manufacturers closely observe the loopholes and try to fill them with their brilliant ideas. You may see a broad set of people shifting from standard vehicles to autonomous vehicles in the coming few years. It is a kind of vehicle that takes away much of the load from the driver's hands and provides specialized systems. The present challenges may seem to be holding back the development of autonomous vehicles, but advancements in technologies will likely overcome all such underlying challenges. In the current scenario, there are few companies, as you saw, who have already released a few autonomous vehicles with notable features. The market is now set up for the number of autonomous vehicles to increase. Sooner or later, you may even find several new algorithms and devices coming into the picture to solve challenges related to autonomous vehicles. Trust issues and social acceptance are two significant points of interest for these vehicles to survive in all countries. In conclusion, you can note that autonomous vehicles are realistic and have a great potential to take over the existing standard or luxurious vehicles.

References

1. Shuttleworth, J. (2019). SAE Standards News: J3016 automated-driving graphic update. *SAE International.*
2. Gupta, Anil. "Machine Learning Algorithms in Autonomous Driving." *IIoT World*, 15 Mar. 2018, Retrieved from https://www.iiot-world.com/artificial-intelligence-ml/machine-learning/machine-learning-algorithms-in-autonomous-driving/, Access Date: 29/10/2021
3. Mohseni, S., Pitale, M., Singh, V., & Wang, Z. (2019). Practical solutions for machine learning safety in autonomous vehicles. arXiv preprint arXiv:1912.09630.
4. Press Release 105/2016: Karlsruhe to Pioneer Autonomous Driving
5. Ziegler, J., Bender, P., Lategahn, H., Schreiber, M., Strauß, T., & Stiller, C. (2014, April). Kartengestütztes automatisiertes fahren auf der bertha-benz-route

von mannheim nach pforzheim. In *Workshop Fahrerassistenzsysteme.* Walting, Germany (Vol. 496).

6. Road traffic injuries, World Health Organization (WHO), 2021, Retrieved from https://www.who.int/news-room/fact-sheets/detail/road-traffic-injuries, Access Date: 29/10/2021
7. Burke, K. (2019, April 15). How Does a Self-Driving Car See? Retrieved from https://blogs.nvidia.com/blog/2019/04/15/how-does-a-self-driving-car-see/, Access Date: 30/10/2021
8. Peter J. Leiss, 2018, "The Functional Components of Autonomous Vehicles", Robson Forensic. Retrieved from https://www.robsonforensic.com/articles/autonomous-vehicles-sensors-expert/, Access Date: 01/11/2021
9. Navarro, P. J., Fernandez, C., Borraz, R., & Alonso, D. (2017). A machine learning approach to pedestrian detection for autonomous vehicles using high-definition 3D range data. *Sensors*, 17(1), 18.
10. Wang, B. (2019, November). Research on Pedestrian Detection Algorithm Based on Image. In *Journal of Physics: Conference Series* (Vol. 1345, No. 6, p. 062023). IOP Publishing.
11. Fleyeh, H., & Dougherty, M. (2005, September). Road and traffic sign detection and recognition. In *Proceedings of the 16th Mini-EURO Conference and 10th Meeting of EWGT* (pp. 644-653).
12. Assidiq, A. A., Khalifa, O. O., Islam, M. R., & Khan, S. (2008, May). Real time lane detection for autonomous vehicles. In *2008 International Conference on Computer and Communication Engineering* (pp. 82-88). IEEE.
13. Bertozzi, M., & Broggi, A. (1998). GOLD: A parallel real-time stereo vision system for generic obstacle and lane detection. *IEEE Transactions on Image Processing*, 7(1), 62-81.
14. Samuel, M., Mohamad, M., Saad, S. M., & Hussein, M. (2018). Development of edge-based lane detection algorithm using image processing. JOIV: *International Journal on Informatics Visualization*, 2(1), 19-22.
15. Canny, J. (1986). A computational approach to edge detection. *IEEE Transactions on Pattern Analysis and Machine Intelligence*, (6), 679-698.
16. Anand, A. (2021, August), "Top Self Driving Car Companies In 2022", Analytic Steps, Retrieved from https://www.analyticssteps.com/blogs/top-self-driving-car-companies-2022, Access Date: 12/11/2021
17. Sangmin, L., Younghoon, K., Hyungu, K., Soon-Kyo, L., Seokhyun, C., Taesu, C., ... & Seoung, B. K. (2020). Intelligent traffic control for autonomous vehicle systems based on machine learning J]. *Expert Syst. Appl*, 144.
18. Tuncali, C. E., Fainekos, G., Prokhorov, D., Ito, H., & Kapinski, J. (2019). Requirements-driven test generation for autonomous vehicles with machine learning components. *IEEE Transactions on Intelligent Vehicles*, 5(2), 265-280.

6

Review on Security Vulnerabilities and Defense Mechanism in Drone Technology

Chaitanya Singh* **and Deepika Chauhan**

Department of Computer Engineering, Swarrnim Startup and Innovation University, Gandhinagar, Gujarat, India

Abstract

Due to the emergence of new technologies such as the Internet of Things, the demand for drones has increased significantly. The concept of the Internet of Drone Things (IoDT) is envisioned as the next generation of drones, which will be equipped with various sensors and technologies such as Cloud Computing, Big Data, and GPS. With its widespread adoption, drones will become more intelligent and useful for monitoring various sectors such as agriculture, government, and industrial. Although they can be used in various industrial sectors, their use has also faced various security concerns. This chapter aims to provide a comprehensive overview of these issues. This paper provides a step-by-step guide on how to protect drones from security issues, including identifying and mitigating known risks. It also discusses various countermeasures that can be used to prevent exploitation of these systems.

Keywords: Autonomous vehicles, drone, wireless communication, intelligent system, cyber attacks, spoofing, DDoS

6.1 Introduction

Unmanned aerial vehicles are cyber-physical systems that integrate various physical and cyber processes. Their main technology components are aircraft structures, propulsion systems, and embedded computing systems. Cybersecurity has become an important concern for drones since they

---

**Corresponding author*: er.chaitanyasingh@gmail.com

Romil Rawat, A. Mary Sowjanya, Syed Imran Patel, Varshali Jaiswal, Imran Khan and Allam Balaram (eds.) Autonomous Vehicles Volume 1: Using Machine Intelligence, (113–132) © 2023 Scrivener Publishing LLC

have unique characteristics such as their dispersed network and physical systems. Due to their unique characteristics, they are vulnerable to exploitation by cyber-attacks [1].

The rise of automated systems has increased the number of civilian and military applications of UAVs. They are known to perform various tasks such as monitoring natural disasters, conducting wildlife surveys, and monitoring weather reports. Due to the nature of their payloads, UAVs are more susceptible to cyber-attacks. There are some security issues that can affect the operation of these unmanned aircraft. Aside from their payloads, control station modules are also vulnerable to exploitation [2]. Communication security is especially important for UAVs since they tend to carry sensitive information that could be easily accessed by adversaries. Some of the most common types of attacks that can affect these systems include denial-of-service, session hijacking, information injection, eavesdropping, man-in-middle attack and GPS spoofing [3].

Section 6.2 discusses the background of the security threats and vulnerabilities that exist in the drones followed by the various author studies. Section 6.3 presents the categorization of attacks to which drones are subjected to. Section 6.4 targets the defense mechanism and countermeasure towards attacks on the drone technology. Finally, section 6.5 concludes the paper.

6.2 Background

A drone is a remote-controlled aircraft that uses GPS and camera-free flight. In the UK, there are more than 80,000 registered flyers/operators of commercial drones. Table 6.1 presents a review of past attacks done on drone technology-based applications.

Yaacoub, J. P. *et al.* wrote that jamming and spoofing techniques can be used to disable a drone, though they can only be performed in certain circumstances. For instance, jamming and spoofing can prevent a drone from recognizing and using the same GNSS signal. However, they can also be used to trick a receiver into thinking that it is located elsewhere [4].

Most of the time, the discussions about smart devices' security and privacy issues are focused on explaining the various types of threats and attacks that are posed by these devices. However, the detailed analysis of these threats can only be presented in a way that shows the attack surface and its various attack methods. Many authors discussed various types of attacks that can be performed by sensors and the various security features that are available in the cloud-based platforms used for the Internet of Things. They also explained how these features are adapted in smart apps.

Siddappaji B. *et al.* discussed the fact that cybersecurity has become an important concern for drones since they are vulnerable to attacks due to their unique design and dispersed physical systems. This paper aims to address the various concerns of drones and provide a comprehensive overview of cyber security [5]. Jean-PaulYaacoub *et al.* analyze the various threats posed by drones and the countermeasures that can be used to prevent them from being attacked. The authors also review the various uses of drones, such as weaponized drones and civilian drones. It also provides a realistic attack scenario that shows how an attack on a drone would work [6]. This review will allow ethical hackers to identify the existing weaknesses of drones and develop new techniques and technologies to prevent them from being attacked.

H. Kim *et al.* introduce a framework for Ai-driven 5G applications, namely, autonomous vehicles and drones. This paper will discuss the various applications of 5G, such as autonomous vehicles and drones. It will also introduce various security threats and solutions that can be used for these new technologies [7]. Due to the increasing popularity of 5G networks, it is widely expected that various autonomous systems and devices will be influenced by the technology. Aside from these, AI is also being widely used to develop new security threats and solutions for various applications. Table 6.1 presents a survey of various security attack and vulnerabilities that exist in the drones.

6.3 Security Threats in Drones

The increasing popularity of drones has raised concerns about their safety and privacy. Although they are very useful in certain scenarios, they can also be misused. This section describes the various types of attacks that an AV may encounter [20]. They are grouped according to the attacker's type, method, target, and motive. Figure 6.1 presents the taxonomy of drone attacks. These attacks are categorized into three categories: electronics attacks, cyber-attacks and physical attacks.

6.3.1 Electronics Attacks

Electronic Attacks are directly connected to the electromagnetic space used for the drone technology communication. These attacks are carried out using electromagnetic spectrum during the transmission and receiving of information via infrared, radio or radar, sensor or sensing signals.

Table 6.1 Survey of security attacks and vulnerabilities in drones.

Cyber security threat	**Security threat**	**Ref.**
Protocol Vulnerabilities	Privacy Leakage	[8, 13]
	Communication Link Security	[8–11]
	Confidentiality Attack	[12]
	Replay Attack	[13, 14]
	De-Authentication Attack	[14, 15]
Sensor Based Attacks	GPS Spoofing	[15, 19]
	Wi-Fi Spoofing	[15]
	Motion Sensor Spoofing	[15]
	UAV Spoofing	[15]
	IP address spoofing	[15]
	Jamming Attacks	[15]
Component or device compromised attacks.	IoT security Threat	[16, 17]
	Device Controlling Threats	
	Data Interception Threats	
Hardware Based Attacks	Password Theft	
	Man-in-Middle Attacks	
Data Base Attacks	Authentication, Integrity, Privacy	
Hijacking Attacks	Authentication, Integrity, Privacy	
Jamming Attacks	Stop Packet Delivery	[15–17]
	Denial of Service	[18]

6.3.1.1 GPS and Communication Jamming Attacks

Jamming is the intentional use of electromagnetic energy to disrupt or deny a communication system's signal transmission. This technique is a threat to GPS and communication systems of UAS. The use of commercially available jammers could allow anyone to carry out this nefarious act. Jamming is a technique that uses high-frequency transmissions to disrupt

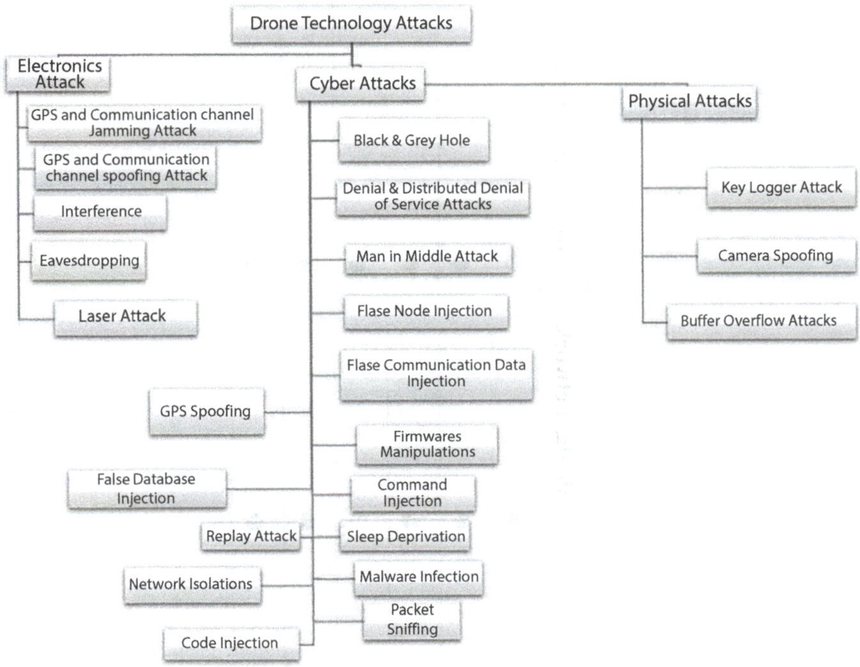

Figure 6.1 Types of attacks in drones.

or block radio communications. This attack can be performed on various types of radio systems such as mobile phones, GPS receivers, and Wi-Fi networks [21]. Table 6.2 depicts the security vulnerabilities in various types of networks.

6.3.1.2 GPS and Communication Spoofing Attacks

Spoofing is a type of attack that occurs when a person or program uses stolen data for illegal gain. It can be performed through various means such as email, cellular, and GPS systems. An attacker can use a fake GPS signal to deceive a receiver [22]. Table 6.3 shows the frequency of occurrence of security vulnerabilities in various types of networks where H represents high, L represents Low and M represents medium.

6.3.1.3 Eavesdropping

Eavesdropping is the interception of the communication signals sent by a transmitter and receiver. An attacker can use this technique to access and

Table 6.2 Security vulnerabilities of various types of networks.

Security issues	**Adhoc networks**	**Sensor networks**	**Mesh network**	**Vehicular network**	**Drone communication**
DoS	●	●	●	●	●
Man in Middle	●	●	●	●	●
GPS Spoofing	○	○	○	●	●
Radar	○	○	○	○	●
Jamming	●	○	●	●	●
Worm Hole	●	●	●	●	●

Table 6.3 Frequency of occurrence of security vulnerabilities in various types of networks.

Security issues	Adhoc networks	Sensor networks	Mesh network	Vehicular network	Drone communication
DoS	L	H	L	M	H
Man in Middle	M	H	L	L	H
GPS Spoofing				M	H
Radar					H
Jamming	M		L	L	H
Worm Hole	H	H	L	L	H

modify the contents of encrypted signals sent by a UAS device. An attack with eavesdropping occurs when a user connects to an unencrypted network and sends sensitive data to a colleague. The unencrypted data is then transported across an open network, which an attacker can exploit through various techniques.

6.3.1.4 Electromagnetic Interference

Electromagnetic interference is a type of attack that can disrupt a UAS communication link. It usually causes it to get degraded and prevent the aircraft from communicating with the ground. Most drones rely on the ground-based communication systems of their UAVs to keep in contact with each other. However, due to the vast number of structures and towers in an area, UAS might encounter electromagnetic interference. This type of interference can severely affect a drone's navigation system. It can also cause it to lose its GPS reception when flying near large structures. In addition, it can affect the communication between the drone and its controller. When a drone is inspecting a cell-tower antenna, it might get hit by the interference. Usually, it returns to its home base to avoid experiencing the damage caused by the high-voltage signal [23].

6.3.1.5 Laser Attacks

Laser attacks are becoming more prevalent, which has prompted the government of New Zealand to draft an amendment bill. Due to the increasing number of laser attacks on aircrafts, it has been concluded that the use of lasers to disable a remote pilot view is a rising threat. Aside from the weapons themselves, state actors can also use them against UAS [24].

6.3.2 Cyber-Attacks

An attack is performed on a communication network or device for the purpose of disrupting or destroying a computing environment or stealing control of it. These types of attacks are done using the cyber space. Unlike physical attacks, cyber-attacks are harder to detect and are often not revealed until after they have been compromised. Attacks on UAS rely on the existing weaknesses of their target systems. This paper will describe the various cyber physical threats that can be encountered against UAS.

6.3.2.1 *Man-in-Middle Attacks*

One of the most dangerous attacks against UAS is an attempt to de-authenticate the aircraft's communication link. This method can allow an adversary to take over the UAS by altering its communication link. In 2016, an attacker posing as a control tower was able to remotely control an aircraft in Australia. The incident led to an investigation by the country's police.

6.3.2.2 *Black Hole and Grey Hole*

A black hole is a network component that draws packets from the other nodes by tricking them into thinking that its route has been established. It can also cause routing protocols to be disturbed by its behavior. On the other hand, Grey hole is a type of attack that can switch from being a normal node to a malicious one. Since it can behave like a normal node, it can't be identified as the intended target. In order to determine the exact route from its origin to its destination, every node checks its routing table [25].

6.3.2.3 *False Node Injection*

The primary function of a drone is to monitor and sense the observed physical phenomena, such as traffic accidents, forest fires, and injured persons. It is also used to report the same phenomena to the same neighbors. Unfortunately, if a malicious node gets into the network, it could cause a serious issue for the monitored drones. An attacker can create a weak communication channel link to transmit false data to a receiver. This can happen through sensor communication links or control links [26].

6.3.2.4 *False Communication Data Injection*

An adversary can compromise a weak communication channel link by sending and receiving false data. This attack can be executed over a variety of sensor communication links. False data injection attacks are usually random and can compromise the integrity of the information gathered from smart meters [27].

6.3.2.5 *Firmware's Manipulations*

CP-embedded systems are targeted by firmware manipulation attacks, which can be done remotely. This attack works by tricking a device into thinking that it is connected to a network.

6.3.2.6 Sleep Deprivation

Sleep deprivation is a type of attack that prevents victims from performing energy-intensive tasks. It causes the battery to consume more energy. Sleep deprivation attacks drain the battery of a device more quickly than it would otherwise. It prevents the device from being used. When the device is idle, its power management software automatically turns it into sleep and low-power standby mode. An attacker can then prevent the device from entering these modes by disabling it.

6.3.2.7 Malware Infection

Malware is a wide range of software that can infect various types of systems. Some of these include viruses, worms, and trojan horses. They can perform various harmful activities, such as spreading viruses and gaining unauthorized access to a network [28].

6.3.2.8 Packet Sniffing

Packet sniffing attacks are usually carried out by an adversary who gains unauthorized access to the targeted network by using an eavesdropping application. This attack is carried out by gathering and analyzing the traffic coming from and through a network. It is done by using tools that monitor and intercept the data packets. There are two types of sniffing, active and passive [29].

6.3.2.8.1 Active Sniffing
Is a type of packet sniffing that uses a switch to monitor the flow of traffic between its ports.

6.3.2.8.2 Passive Sniffing
Is different from Active Sniffing. It uses a network segment to monitor the traffic coming from and through a non-switched network.

6.3.2.9 False Database Injection

The term false data injection attack was first used in the smart grid industry. It refers to an attack where an attacker steals a sensor's readings in order to trick a system into generating erroneous values. Though the concept of Data Infiltration Attacks (DIAs) originated from the smart grid industry, they can be applicable to any Internet-connected environment. This is a technique that an unauthorized individual or group of hackers uses to take over an Unmanned Aerial Vehicle (UAV). Usually, an attacker uses their own sensors to inject fake data into the UAVs to take over their operations. Researchers analyzed the movement of the UAVs using intelligence sensors. They then implemented a neural network framework to improve its accuracy and time efficiency. To detect the fake information, a threshold technique was used. Although the use of threshold techniques is not ideal for detecting fake data, it is still important that the UAVs are trained to prevent getting exposed to this type of information. Also, researchers are still trying to find a way to secure the data transmission over a wireless network.

6.3.2.10 Replay Attack

A command can be captured and relayed to another system for further investigation or manipulation. This can lead to an attack on a UAS system that demands a command or action to be executed. A replay attack is a type of security attack that uses an attacker to capture the traffic sent over a network. The attacker then sends the message to the intended recipient, which is typically an authenticated message.

6.3.2.11 Network Isolations

An adversary can prevent a large network from communicating with other systems by creating a geographic area that is isolated from the rest of the network.

6.3.2.12 Code Injection

Code injection attacks are usually performed in order to modify the behavior of a particular software component or system. They can expose various bugs in the system and expose sensitive information.

6.3.3 Physical Attacks

An attack performed in a physical space is considered physical security. These types of attack occur on a physical device connected in the system.

6.3.3.1 Key Logger Attacks

This is a type of virus which attacks by analyzing the keystrokes to gain access or confidential information.

6.3.3.2 *Camera Spoofing*

A buffer overflow occurs when the amount of data that can be accessed exceeds the capacity of the memory buffer. This overflow can happen even if the program has allocated enough space for the buffer. It can also cause a program to execute unpredictably if it tries to read the data past the buffer boundary.

6.4 Defense Mechanism and Countermeasure Against Attacks

This section presents the defense mechanism and countermeasures to be taken for the protection of drones against various attacks. This section describes the defensive countermeasures and various techniques used for the determination and prevention of the attacks in drones. Table 6.4 represent the vulnerabilities and their countermeasures in the protocol used by the drones at various layers.

6.4.1 Defense Techniques for GPS Spoofing

There are various techniques that can be used to prevent a hacker from launching a GPS spoofing attack. D. Borio and C. Gioia state that one of these is by jamming the signals, which are typically generated by the GPS receiver. Although this technique can detect the high-intensity signals generated by a spoofing attack, it can also prevent them from being replicated [32].

Due to the nature of GPS signals, they can be easily spoofed. A. J. Kerns *et al.* introduce the use of asymmetric cryptography to prevent the exploitation of GPS signals. Another technique that can help prevent the exploitation of GPS signals is the use of a supervised machine learning approach [33]. M. R. Manesh *et al.* present a method that uses the features of original GPS signals to generate a machine learning algorithm that can identify fake packets of GPS signals. The algorithm is also used to prevent the exploitation of the original GPS data [34].

6.4.2 Defense Technique for Man-in-Middle Attacks

Various authors proposed various techniques to prevent GPS spoofing attack. The first one is by detecting and avoiding a man-in-middle attack before sending data signals. Another one is by encryption at the hardware level. There are various techniques to prevent man-in-middle attacks.

Table 6.4 Vulnerabilities and their countermeasures in drone technology protocols.

Category	Technique	Detail	Countermeasures
Global Positioning System (GPS)	GPS spoofing attack detection [30].	Comparison of techniques using three (Euclidian distance, Angle Distance and histogram of oriented distance) parameters.	High accuracy increases the frequency of detection of attacks.
	Context based detection of GNNS positioning spoofing [31].	Measures integrity, correctness and consistency-based detection schemes for the spoofing.	Depend on the quality of context.
	GPS spoofing attack detection [30].	Detection using nonlinear fusion of sub band classifier.	
	GPS spoofing attack detection [36].	Detection on single and multiple dynamic attack.	Uncertainty of attack decreases the probability of decentralized spoofing attacks detection.
	GPS spoofing attack detection [37].	Location-based spoofing attack detection.	proposed a method to detect location spoofing attacks, which involve taking advantage of the GPS and accelerometer features of most mobile devices.
False Data Injection	Deep Learning-based FDI attack detection [38].	Detection of FDIA using data and learned features.	Convolution Network will generate better accuracy.
	CNN based FDI attack detection [39].	Detection of FDIA on the basis of temporal behavior.	Inclusion of pooling and deeper network increase accuracy of detection.

(Continued)

Table 6.4 Vulnerabilities and their countermeasures in drone technology protocols. (*Continued*)

Category	Technique	Detail	Countermeasures
Wi-Fi Insecurity	Micro air vehicle communication protocol vulnerability analysis [43].	Due to time inefficiency MAV Link protocol is a unable to perform encryption.	
Jamming Attack	Preventing UAVs from jamming attacks [42].	Detection and elimination of artificial noise inclusion by the attacker.	Watermark technique enabled plaintext inclusion make the method effective.
	Physical layer security improvement [43].	Artificial noise inclusion and height of drone is fixed.	Worked with variable parameters with no fixed height and position.
Control System Security Threats	Detection of hardware failure performance [45].	Infeasible to detect spoofing attack. Only use recursive least square method for the detection of parameter affecting the control system security attacks.	Jamming technique is used to identify undiscovered spoofing attacks.
Fuzzing Attack	Detection of attacks and vulnerabilities [46].	Three different mechanisms are presented which leads to overhead.	Optimization of algorithm is required.
Malicious Drone Detection	Malicious drone detection using machine learning techniques [47].	Compatibly less accurate	Need of expansion of feature vectors.
	Detection of authentication and authorization vulnerabilities in drone [48].	Fingerprint-based technique is presented.	Instead of fingerprint feature vector needs to be expanded using image processing, signal, etc.

Strong WEP/WAP Encryptions: Having the proper encryption on wireless access points can prevent unauthorized users from accessing your network. A weak encryption can allow an attacker to easily modify the settings of the network.

Strong Router Login Credentials: It is also important to change your default router login credentials. If your credentials are compromised, an attacker can easily modify your DNS servers and infect your network.

Virtual Private Network: A virtual private network is a type of network that enables you to secure an environment within a local area network. It uses key-based encryption to prevent unauthorized access.

Force HTTPS: HTTPS is a commonly used HTTP format to communicate with websites. It can be secured by using a public-private key exchange (PSB) instead of a HTTP alternative.

6.4.3 Defense against Keylogger Attacks

There are many solutions for the keylogging attack, but two main approaches are commonly used to mitigate it. One is by installing an anti-virus software which can detect the presence of key loggers in a system. Another strategy is to prevent the exploitation of sensitive information by examining the applications that can access it. Protect yourself from unauthorized access by setting up firewalls. In most cases, a keylogger has to send sensitive information to a third party in order to perform its duties. One way to do this is by creating a form-filler that automatically fills in all the personal information that you enter. The keyloggers only see the scrambled keys. Be aware that anti-virus programs can flag these types of programs as malware, but they can also ignore any false positive alarms. Never use a one-time password (OTP) or an alternative method of authentication. Two-factor authentication is also used to prevent exploitation. A website that uses this exploit could install a keylogger on your PC. Phishing techniques are used to confuse users and spread infections. Never send or receive emails with links to legitimate websites.

6.4.4 Defense against Camera Spoofing Attacks

A different approach is presented that involves analyzing the movement of a camera. D. Davidson *et al.* proposed a RANSAC algorithm and then used it to identify the most appropriate hypothesis for the frame [35]. Petit *et al.* offer several suggestions to prevent camera-based attacks. They suggest incorporating multiple cameras in order to provide the same view and to prevent infrared light interference. They also suggest using photochromic lenses to filter different wavelengths of light. A method that uses ground-truth and early fusion techniques to monitor objects is also proposed [40].

6.4.5 Defense against Buffer Overflow Attacks

An algorithm that can determine the length of the input characters can be used to prevent attacks that are related to zero-day exploits. Another strategy that can help minimize this type of risk is implementing a machine learning approach to detect inauthentic behavior [41].

6.4.6 Defense against Jamming Attack

An approach to prevent jamming of an Unmanned Aircraft System is proposed through a hardware technique known as the sandbox. This method consists of monitoring the signals from the system and isolation of the non-important component of the UAV. A different approach is to use an antenna array to detect jamming signals [44].

6.5 Conclusion

Due to the increasing number of security issues related to drones, various techniques have been proposed to minimize these risks. However, many of these solutions are not yet effective enough to prevent attacks against drones' components. Instead of relying on GPS, we should develop more effective techniques that can detect and prevent spoofing and jamming. Another challenging part of this process is developing a more secure communication protocol. After conducting a survey, we discovered that the current security protocols for UAVs are not secure enough. The researchers are expected to step up their efforts in developing new techniques and solutions that will make the drones more secure.

References

1. K. Mabodi, M. Yusefi, S. Zandiyan, L. Irankhah, and R. Fotohi, "Multi-level trust-based intelligence schema for securing of Internet of Things (IoT) against security threats using cryptographic authentication," *J. Supercomput.*, vol. 76, no. 6, pp. 1–26, 2020.
2. T. Fox-Brewster. Maldrone: Watch Malware That Wants to Spread its Wings Kill a Drone Mid-Flight. *Forbes*. Accessed: Feb. 27, 2021. [Online]. Available:

https://www.forbes.com/sites/thomasbrewster/2015/01/27/malware-takes-downdrone/?sh=65e60af44c92.

3. H. Zhu, M. L. Cummings, M. Elfar, Z. Wang, and M. Pajic, "Operator strategy model development in UAV hacking detection," *IEEE Trans. Human-Mach. Syst.*, vol. 49, no. 6, pp. 540–549, Dec. 2019.
4. Yaacoub, J. P., Noura, H., Salman, O., & Chehab, A. (2020). Security analysis of drones systems: Attacks, limitations, and recommendations. *Internet of Things*, 11, 100218.
5. Siddappaji B., Akhilesh K.B. (2020) Role of Cyber Security in Drone Technology. In: Akhilesh K., Möller D. (eds.) *Smart Technologies*. Springer, Singapore. https://doi.org/10.1007/978-981-13-7139-4_13.
6. Jean-Paul Yaacoub, Hassan Noura, Ola Salman, Ali Chehab, Security analysis of drones systems: Attacks, limitations, and recommendations, *Internet of Things*, Volume 11, 2020, 100218, ISSN 2542-6605, https://doi.org/10.1016/j.iot.2020.100218.
7. H. Kim, J. Ben-Othman, L. Mokdad, J. Son and C. Li, "Research Challenges and Security Threats to AI-Driven 5G Virtual Emotion Applications Using Autonomous Vehicles, Drones, and Smart Devices," in *IEEE Network*, vol. 34, no. 6, pp. 288-294, November/December 2020, doi: 10.1109/MNET.011.2000245.
8. Vattapparamban, E.; Güvenç, I.; Yurekli, A.I.; Akkaya, K.; Uluagaç, S. Drones for smart cities: Issues in cybersecurity, privacy, and public safety. In *Proceedings of the 2016 international wireless communications and mobile computing conference (IWCMC), Paphos, Cyprus, 5–9 September 2016;* IEEE: Piscataway, NJ, USA, 2016; pp. 216–221.
9. Nayyar, A.; Nguyen, B.L.; Nguyen, N.G. The internet of drone things (IoDT): Future envision of smart drones. In *First International Conference on Sustainable Technologies for Computational Intelligence*; Springer: Berlin/Heidelberg, Germany, 2020; pp. 563–580.
10. Ozmen, M.O.; Behnia, R.; Yavuz, A.A. IoD-crypt: A lightweight cryptographic framework for Internet of drones. arXiv 2019, arXiv:1904.06829.
11. Bertino, E. Data Security and Privacy in the IoT. In *Proceedings of the 19th International Conference on Extending Database Technology (EDBT), Bordeaux, France, 15–18 March 2016*; Volume 2016, pp. 1–3.
12. Lin, C.; He, D.; Kumar, N.; Choo, K.K.R.; Vinel, A.; Huang, X. Security and privacy for the internet of drones: Challenges andsolutions. *IEEE Commun. Mag.* 2018, 56, 64–69.
13. Rodday, N. Hacking a professional drone. *Black Hat Asia* 2016, 2016, 1–2.
14. Highnam, K.; Angstadt, K.; Leach, K.; Weimer, W.; Paulos, A.; Hurley, P. An uncrewed aerial vehicle attack scenario and trustworthy repair architecture. In *Proceedings of the 2016 46th Annual IEEE/IFIP International Conference on Dependable Systems and Networks Workshop (DSN-W), Toulouse, France, 28 June–1 July 2016*; IEEE: Piscataway, NJ, USA, 2016; pp. 222–225.

15. N. Gershenfeld, R. Krikorian, and D. Cohen, "The Internet of things," *Scientific American*, October, 2004.
16. M. A. Shahid, U. Akram, M. M. A. Shahid, A. Samad, M. F. Mushtaq and R. Majeed, "A Systematic Approach Towards Compromising Remote Site HTTPS Traffic Using Open Source Tools," *IEEE 23rd International Multitopic Conference (INMIC)*, 2020.
17. G. Biamino, "Semantic model for socially aware objects," *Advances in Internet of Things*, vol. 2, pp. 47-55, 2012.
18. I. Ungurean, N. C. Gaitan, and V. G. Gaitan, "An IoT architecture for things from industrial environment," in *Proc. 10th Int. Comm.*, 2014, pp.1–4.
19. C.A. Thiels, J.M. Aho, S.P. Zietlow and D.H. Jenkins, "Use of unmanned aerial vehicles for medical product transport", *Air Med. J.* 34(2) (2015) 104–108.
20. L. Mokdad, J. Ben-Othman, A. T. Nguyen, Djavan: Detecting jamming attacks 840 in vehicle ad hoc networks, *Performance Evaluation* 87 (2015) 47–59. doi:10.1016/j.peva.2015.01.003.
21. A. F. Van Niekerk and L. Combrinck, "The use of civilian-type GPS receivers by the military and their vulnerability to jamming," *S. Afr. J. Sci.*, vol. 108, no. 5–6, pp. 4–7, 2012, doi: 10.4102/sajs.v108i5/6.749.
22. E. Horton and P. Ranganathan, "Development of a GPS spoofing apparatus to attack a DJI Matrice 100 Quadcopter," *J. Glob. Position.Syst.*, vol. 16, no. 1, 2018, doi: 10.1186/s41445-018-0018-3.
23. da Silva, M. F., Honório, L. M., Marcato, A. L. M., Vidal, V. F., & Santos, M. F. (2020). Unmanned aerial vehicle for transmission line inspection using an extended Kalman filter with colored electromagnetic interference. *ISA Transactions*, 100, 322–333.
24. Schleijpen, R., van Binsbergen, S., Geljon, M., Meuken, D., Deiana, D., & van Leeuwen, B. (2020, September). 30kW laser experiments against drones. In *Technologies for Optical Countermeasures XVII; and High-Power Lasers: Technology and Systems, Platforms, Effects IV* (Vol. 11539, p. 115390A). International Society for Optics and Photonics.
25. He, D., Chan, S., & Guizani, M. (2017). Drone-assisted public safety networks: The security aspect. *IEEE Communications Magazine*, 55(8), 218-223.
26. Gu, Y., Yu, X., Guo, K., Qiao, J., & Guo, L. (2021). Detection, estimation, and compensation of false data injection attack for UAVs. *Information Sciences*, 546, 723–741.
27. Zuo, Z., Cao, X., & Wang, Y. (2020). Security control of multi-agent systems under false data injection attacks. *Neurocomputing*, 404, 240–246.
28. Yahuza, M., Idris, M. Y. I., Ahmedy, I. B., Wahab, A. W. A., Nandy, T., Noor, N. M., & Bala, A. (2021). Internet of Drones Security and Privacy Issues: Taxonomy and Open Challenges. *IEEE Access*, 9, 57243–57270.
29. Guo, R., Wang, B., & Weng, J. (2020, April). Vulnerabilities and Attacks of UAV Cyber Physical Systems. In *Proceedings of the 2020 International Conference on Computing, Networks and Internet of Things* (pp. 8–12).

30. Varshosaz, M., Afary, A., Mojaradi, B., Saadatseresht, M., & Ghanbari Parmehr, E. (2020). Spoofing detection of civilian uavs using visual odometry. *ISPRS International Journal of Geo-Information*, 9(1), 6.
31. Formaggio F., Ceccato S., Laurenti N., Tomasin S. (2021) Context-Based Detection of GNSS Position Spoofing for Smartphones. In: Paiva S. (eds.) *Precision Positioning with Commercial Smartphones in Urban Environments.* EAI/Springer Innovations in Communication and Computing. Springer, Cham. https://doi.org/10.1007/978-3-030-71288-4_1.
32. D. Borio and C. Gioia, "Real-time jamming detection using the sum-of-squares paradigm", in *Proc. 2015 Int. Conf. Localization GNSS ICL-GNSS 2015*, 2015.
33. A. J. Kerns, K. D. Wesson and T. E. Humphreys, *A Blueprint for Civil GPS Navigation Message Authentication*, pp. 262–269.
34. M. R. Manesh, J. Kenney, W. C. Hu, V. K. Devabhaktuni and N. Kaabouch, "Detection of GPS Spoofing Attacks on Unmanned Aerial Systems", in *2019 16th IEEE Annu. Consum. Commun. Netw. Conf. CCNC 2019*, pp. 1–6, 2019.
35. D. Davidson, H. Wu, R. Jellinek, T. Ristenpart, and V. Singh, "Controlling UAVs with Sensor Input Spoofing Attacks," *10th USENIX Work.Offensive Technol. (WOOT 16)*, 2016.
36. S. Siamak, M. Dehghani and M. Mohammadi, "Dynamic GPS Spoofing Attack Detection, Localization, and Measurement Correction Exploiting PMU and SCADA," in *IEEE Systems Journal*, vol. 15, no. 2, pp. 2531–2540, June 2021, doi: 10.1109/JSYST.2020.3001016.
37. Wong, S. K., & Yiu, S. M. (2020). Location spoofing attack detection with pre-installed sensors in mobile devices. *J. Wirel. Mob. Networks Ubiquitous Comput. Dependable Appl.*, 11(4), 16–30.
38. Umar S, Felemban M. Rule-Based Detection of False Data Injections Attacks against Optimal Power Flow in Power Systems. *Sensors*. 2021; 21(7):2478. https://doi.org/10.3390/s21072478.
39. Mode, G. R., Calyam, P., & Hoque, K. A. (2020, April). Impact of false data injection attacks on deep learning enabled predictive analytics. In *NOMS 2020-2020 IEEE/IFIP Network Operations and Management Symposium* (pp. 1–7). IEEE.
40. J. Petit, S. Bas, M. Feiri, and F. Kargl, "Remote attacks on automated vehicles sensors: experiments on camera and lidar," in *Black HatEurope*, 2015.
41. Dey, V., Pudi, V., Chattopadhyay, A., & Elovici, Y. (2018, January). Security vulnerabilities of unmanned aerial vehicles and countermeasures: An experimental study. In *2018 31st International Conference on VLSI Design and 2018 17th International Conference on Embedded Systems (VLSID)* (pp. 398–403). IEEE.
42. C. Liu, J. Lee, and T. Q. S. Quek, "Safeguarding UAV communications against full-duplex active eavesdropper," *IEEETrans. Wireless Commun.*, vol. 18, no. 6, pp. 2919–2931, Jun. 2019.

43. Y. Zhou, P.L. Yeoh, H. Chen, Y. Li, R. Schober, L. Zhuo, and B. Vucetic, "Improving physical layer security via a UAV friendly jammer for unknown eavesdropper location," *IEEE Trans. Veh. Technol.*, vol. 67, no. 11, pp. 11280–11284, Nov. 2018.
44. Y.-M. Kwon, J. Yu, B.-M. Cho, Y. Eun, and K.-J. Park, "Empirical analysis of MAVLink protocol vulnerability for attacking unmanned aerial vehicles," *IEEE Access*, vol. 6, pp. 43203–43212, 2018.
45. Z. Birnbaum, A. Dolgikh, V. Skormin, E. O'Brien, D. Muller, and C. Stracquodaine, "Unmanned aerial vehicle security using recursive parameter estimation," *J. Intell. Robotic Syst.*, vol. 84, nos. 1–4, pp. 107–120, Dec. 2016.
46. M. Hooper, Y. Tian, R. Zhou, B. Cao, A. P. Lauf, L. Watkins, W. H. Robinson, and W. Alexis, "Securing commercial WiFi-based UAVs from common security attacks," in *Proc. MILCOM IEEE Mil. Commun. Conf.*, Nov. 2016, pp. 1213–1218.
47. A. Alipour-Fanid, M. Dabaghchian, N. Wang, P. Wang, L. Zhao, and K. Zeng, "Machine learning-based delay-aware UAV detection over encrypted Wi-Fi traffic," in *Proc. IEEE Conf. Commun. Netw. Secur. (CNS)*, Jun. 2019, pp. 1–7.
48. I. Bisio, C. Garibotto, F. Lavagetto, A. Sciarrone, and S. Zappatore, "Blind detection: Advanced techniques for WiFi-based drone surveillance," *IEEE Trans. Veh. Technol.*, vol.68, no.1, pp. 938–946, Jan. 2019.

7

Review of IoT-Based Smart City and Smart Homes Security Standards in Smart Cities and Home Automation

Dnyaneshwar Vitthal Kudande, Chaitanya Singh* and Deepika Chauhan

Department of Computer Engineering, Swarrnim Startup and Innovation University, Gandhinagar, India

Abstract

In the recent era there has been widespread development of Internet of Things (IoT) applications like smart cities and smart homes. Within this there are different security attacks and threats at different layers of IoT-used devices and structures that may impact smart city and smart homes performance. For assessing the security of IoT-based frameworks and gadgets used in smart cities and smart homes it is becoming vital to utilize the right control measures. Along with these, security is the main fundamental issue that relies on the IoT devices. This paper provides an outline of security threats and preventive measures in smart cities and smart homes.

Keywords: IoT, smart city and home, security threats, blockchain, machine learning

7.1 Introduction

Internet of Things (IoT) based smart applications are becoming popular emerging methodologies among the stakeholders. The term smart applications was created from the possibility of pervasive computing or actual word that contains various sensors, regulators, actuators and other computational components that cooperate with one another to make people's lives more convenient and efficient [1].

**Corresponding author*: er.chaitanyasingh@gmail.com

Romil Rawat, A. Mary Sowjanya, Syed Imran Patel, Varshali Jaiswal, Imran Khan and Allam Balaram (eds.) Autonomous Vehicles Volume 1: Using Machine Intelligence, (133–148) © 2023 Scrivener Publishing LLC

Ikrissi *et al.* (2020) present the idea of Internet of Things (IoT) which states that it provides sensing acting and large data processing capability with security as value-added services in the field of smart cities and home automation [2].

As per the theory given by H. Lin and N. W. Bergmann (2016), IoT has turned into a key empowering agent of development and achievement in a wide scope of fields including IoT-based smart applications. IoT has advance capability of creating other IoT-based smart applications, which permits people to control smart gadgets and machines remotely by utilizing PCs, cell phones, or tablets via the World Wide Web [3].

Since IoT-enabled applications impact our life as we are directly or indirectly connected to smart devices used in a smart city and smart home, they are subject to various privacy and security issues [4].

The smart applications experience several security challenges like DDOS attack, eavesdropping attacks, sniffing attack, SQL injection, etc. IoT-based smart applications, correspondence, and computation assets are utilized in smart cities and smart homes to enhance the personal satisfaction that can be accessible to metropolitan residents. It also handles residents' smart data in regions like smart agriculture, smart traffic, and smart transportation frameworks [5].

A conventional definition exists for smart applications in the IoT that highlights smart devices used on the geographic area and dictated by utilizing a dissecting framework [6]. Sensor administrations for the assortment of individual information can be utilized for a few activities, for example, concerning the checking of cyclists, vehicles, and public parking garages.

Further organization of this paper is done in the following manner: Section 7.2 covers the overview and motivation of IoT with regard to smart applications areas. Section 7.3 presents an existing literature review of the research of various authors. In section 7.4, we present some IoT security threats and preventive measures in smart environments. More exhaustively, section 7.5 focuses on primary solutions related to these security threats. Lastly, section 7.6 discusses the conclusion and proposes future exploration headings.

7.2 Overview and Motivation

This survey paper was inspired by the understanding that conventional security norms and evaluation systems implied for use in non-IoT conditions are totally different and many of them may not straightforwardly address the requirements of IoT-based smart cities and smart homes. This paper accordingly explores the possibilities that conventional security principles and appraisal systems have in tending to IoT-based smart cities and smart homes security worries by uncovering their essential spaces of concentration just as the cutting edge and foundation of the area.

The main purpose of this paper is to present different IoT security threats in smart cities and smart homes. Also, it shows some solutions to IoT-based smart cities and smart homes architecture, that can be divided into four layers: i) Sensor layer, ii) Network layer, iii) Platform layer, and iv) Application layer (Figure 7.1).

(i) Sensor Layer: This is also called as physical layer, as it collects information from the different devices like sensors, actuators, RFID, etc.

(ii) Network Layer: Network layer consists of data link protocols like Wi-Fi, 4G, 5G, ZigBee, etc., and network protocols like IPV4, IPV6, etc. The main purpose of this layer is to access information from sensor layer and transmit it to the platform layer.

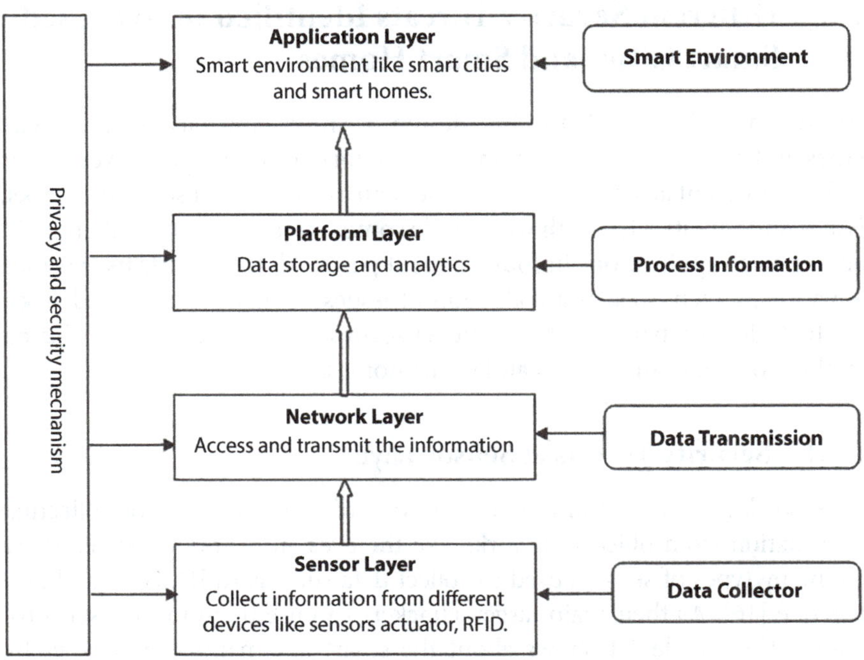

Figure 7.1 The four layer IoT architecture in smart environment.

(iii) Platform Layer: Platform layer consists of different database like SQLite, MongoDB, etc. and data analytics processing. The main purpose of this layer is to provide data storage and data analytics functionality.

(iv) Application Layer: Application layer means actual smart cities and smart homes.

7.3 Existing Research Work

In the literature, a few security principles, evaluation systems, and extraordinary disseminations on security methodology exist which can be used in different conditions (e.g., network security, web security, applications security, media transmission among different regions). Be that as it may, these security principles and evaluation systems were principally planned in view of explicit application conditions, thus various advances or cycles for various conditions are involved as featured later in section 7.4. Analysts in the IoT area have additionally proposed various methodologies and procedures to address diverse IoT lacks and structures, the premise of the current exploration work in this part.

7.4 Different Security Threats Identified in IoT-Used Smart Cities and Smart Homes

As shown in Figure 7.1, the IoT design of smart applications like smart cities and smart homes are involved in the form of four layers. Every layer utilizes different advances, which present an assortment of security attacks, issues and threats. Herewith we include some unique security challenges in an IoT-used applications in four different parts. All these parts use various advances, which present a collection of issues, security attacks and risks. In the following part we present novel security challenges in an IoT-used applications like smart cities and smart homes.

7.4.1 Security Threats at Sensor Layer

A sensor layer is a part of the human body that's responsible for collecting information from objects. It works like the eyes and ears of people. There are many types of sensors used to collect data such as RFID, GPS, and 2-D barcode [16]. As their main target, attackers can use them to steal sensitive information. Table 7.1 shows about the smart automation approaches by highlighting current research work and its main focus regions.

Table 7.1 Current research work and its main focus regions.

Different proposed frameworks	Main focusing regions		
	Smart city and homes	**Cloud-based**	**Health-care based**
A lightweight IoT-based security framework for Inventory automation using wireless sensor network [7].	√	√	√
An effective and efficient risk assessment framework for cloud service provider [8].	√	√	
An enhanced security framework for smart devices in smart city and smart home environment [9].	√		
A framework for evaluating security risks associated with technology used at home [10].	√		
An assessment framework to evaluate security features of IoT-based equipment [11].	√		√
IoT-based integrated home security and monitoring system [12].	√		
A secure and compliant continuous assessment framework for evaluating the security and compliance level of cloud service [13].	√	√	

7.4.1.1 Eavesdropping Attacks

Eavesdropping attacks are very old and are usually carried out by people who are listening to network communications. They can obtain sensitive information such as routing updates and node identification numbers. Fortunately, there are a number of techniques that can prevent eavesdropping attacks without affecting the operation of the network.

The cryptographic primitives of an Internet of Things network are built out of a single block cipher and are commonly reused. The presence of a common state among the communicating parties enables the use of low-cost code reuse. Hassijja *et al.* present a strategy that enables the re-keying and selective distribution of key to individual sensors. This study proposed a protocol that supports in-network processing and provides security properties that are similar to those of pair wise keys. Another paper proposes a method that allows sensors to secure connectivity between them [14].

7.4.1.2 Node Capturing Attacks

An attacker with full control over a key node can easily expose sensitive information. For example, if the data collected by a sensor is stored in memory, it could be compromised [15].

7.4.1.3 Sleep Deprivation Attacks

The concept of the sleep deprivation attack is a proposed method of attacking a battery-powered computing device that has a low power sleep mode. The attacker tries to interact with the victim in a way that appears to be legitimate, but the objective is to keep the victim's sleep mode active. Krishnaswami's approach measured the difference between the average Power consumption of devices while they were idle and when they were attacked. However, we find his approach to be overly simplistic. An adversary would typically target devices that are designed to consume large amounts of energy [14].

7.4.1.4 Malicious Code Injection Attacks

Code injection is a type of attack that occurs when an attacker uses an unauthenticated input in a web application. It can be executed by a server-side interpreter. Most web applications that rely on unauthenticated input are vulnerable to this type of attack. Remote code execution is a type of attack that can be performed by an attacker through an unauthenticated web application. The issue here is that when a user inputs data, the application processes it and can store it in various ways. This includes not only direct input, but also input from other sources such as cookies and web beacons.

7.4.2 Security Threats at Network Layer

Secure network is the key to success in any network environment. In this layer it provides the gateway for servers where all applications reside. This layer is based upon crucial associations like correspondence networks of internet and sensor associations. Its fundamental task is to provide the data connectivity between sensor layer and platform layer [17].

The critical association layer security issues are described below.

7.4.2.1 Distributed Denial of Service (DDOS) Attack

In an IoT association, this attack means to disturb server availability through a flood of imitated requests from dispersed IoT contraptions on the correspondence channel [18]. The association layer of the IoT is vulnerable against them on account of the unpredictability and heterogeneity of IoT associations. Various IoT contraptions used in IoT applications are not securely organized, making them clear targets for DDoS attacks on target servers.

7.4.2.2 Sniffing Attack

It is an attack that remembers tuning for on network packages. This attack is amazingly typical in wired and far off associations, and it hazards correspondence mystery. This attack can be finished by an aggressor using a compromised device or by clearly getting groups from the common medium. Somewhat topological information, directing information, and data content may be gotten from sniffed packages [19].

7.4.2.3 Routing Attack

A routing table poisoning attack occurs when a maliciously altered routing table is detected by a router. It can cause severe damage to the network. Attacks on Network Resources are grouped into three categories: i) attacks on the network resources, ii) and attacks on the network topology [14].

7.4.2.3.1 Selectively Forwarding Attacks

This attack occurs when a packet is selectively forwarded. The goal of this attack is to disrupt the routing paths and filter the traffic. A good technique to prevent this type of attack is by implementing an IDS solution. Attacks that can alter the routing information of a network are known to cause RPL self-healing to fail. They can also expose falsified routes that can lead to network sub-optimization. This attack can only be performed when the routing mode is enabled.

7.4.2.3.2 Sinkhole Attack

In sinkhole attacks, an attacker tries to attract nearby nodes to a certain path. This attack can be very powerful if combined with another attack. To prevent this type of attack, an IDS solution is used. The root process generates a hash value by choosing a random value and then broadcasts it in a DIO message. Each node stores the hash value and the number of hops in its path. After the root node broadcasts a random number, the other nodes can then verify its parent. If the DIO message contains an error, the system can then add the parent.

7.4.2.3.3 Wormhole Attack

A wormhole attack can disrupt the traffic flow and the network's topology. This attack can be performed by creating a tunnel between two attackers and taking advantage of the existing traffic. One of the techniques that can prevent this attack is implementing the tree authentication method.

7.4.2.4 Traffic Examination Attacks

In this type of attack the aggressor initially secures network information by using port analyzing or group sniffers then examining the association traffic from customer devices in solicitation to get required information about customers. This is a segregated attack that is difficult to recognize [20].

7.4.3 Security Threats at Platform Layer

It fills in as a middle person layer between the association and application layers. Right when IoT devices interface and pass on with one another, they produce a collection of organizations. The Stage layer's ability is to give solid figuring limit capacities, which suggest that it can store the information of the lower layer in data bases and moreover recuperate process, register, besides pick subject to the computational results [21]. The following different types of attacks happened in the middleware part.

7.4.3.1 SQL Injection

In data-driven applications, using code injection method this type of attack can be done by an attacker. In an IoT-based database system an attacker may use malicious SQL statements for performing SQL injection attack. In many cases the attacker may change or delete the field in the database that may cause persistent changes in the given database [22].

7.4.3.2 Cloud Malware Injection

This type of attack is performed when a user opens an account under a cloud vendor, and the cloud vendor takes control of a user's information. For this the hacker may add infected service module to a virtual machine instance. The most common type of malware injection attack is cross site scripting attacks or SQL injection attack [21].

7.4.3.3 Storage Attacks

In dynamic storage condition the large amount of data contains the customer's irreplaceable data which can be taken care of by the cloud devices. The duplication of data combined with the openness of data and different types of user brings about a greater surface area for attacks [14].

7.4.3.4 Side Channel Attacks

In this type of attack the attacker must observe the information by identifying revealed side features of the algorithmic execution which is similar to control different usage that taking care of timing and related sounds, etc. This type of attack could happen due to a shortfall of secure type for dealing with and taking care of IoT data at that instance taking care of decoded data in the IoT object or cloud [23].

7.4.4 Security Threats at Application Layer

Application layer is the uppermost layer of the latest IoT design, which is liable to give clients smart and intelligent applications that are reliant upon their necessities. Herewith we consider the different threats at application layer in smart cities and smart homes.

7.4.4.1 Sniffing Attack

In network security terminology, sniffing attacks means interception of information by capturing network traffic using a specific tool called as packet sniffer. When data packets are not encrypted over the network then it can be read by sniffers; the attacker presents a sniffer application into the IoT structure to force an attack on it, which could gain network information, occurring in a spoiled system and allowing the assailant permission to access private customer data if sufficient security shows are not set up to hinder it [25].

7.4.4.2 *Reprogram Attack*

In specific conditions, an association programming structure can be used to remotely remake IoT objects. If the most widely recognized method of composing PC programs isn't gotten, an assailant could use it to accept accountability for a gigantic piece of the IoT association [24].

7.4.4.3 *Data Theft*

IoT-used devices nowadays become a main target of cybercriminals. In this system cybercriminals can get easy access to private information due to lack of layer-wise security protocols. The different advisory mostly used an untrustworthy method of course a couple of data dealing with estimations to take customers' restricted information, which may achieve shocking mischief.

7.4.4.4 *Malicious Script Attack*

This type of attack is also called as a DDOS attack. In this attack, the programmer signs into the framework and professes to be a verified client to upset the organization's typical activity and forestall genuine clients from utilizing the administration's [25]. This is one of the framework's generally serious and risky issues. The most regularly involved components in the topmost part to ensure information protection, validation, encryption and trustworthiness.

7.5 Security Solutions For IoT-Based Environment in Smart Cities and Smart Homes

Herewith we present a couple of preventive measures for overseeing security attacks in the IoT engineering layers. Herewith we provide some essential security measures to prevent different threats in smart homes and smart cities.

7.5.1 Blockchain

Blockchain is a shared conveyed unified and public decentralized technique for taking care of plans, arrangements and trades across various PCs. It is fundamentally a chain of squares wherein mechanized data is taken care of in a public informational index. Blockchain is arranged expressly for computerized monetary forms like Bitcoin and Ethereum. Blockchain advancement can be a key to controlling, regulating and specifically, settling various IoT-used assets of security. Also, the entire trades delivered utilizing an IoT-used contraption have been enrolled in the blockchain dispersed record that may be continued in an ensured manner [15]. Besides, by giving reasoning and de-concentrated approval rules, Blockchain brilliant arrangements provide the assurance to multiparty and single and affirmation in the IoT-used device [15].

7.5.2 Lightweight Cryptography

Lightweight cryptography provides an encryption technique which features little computational complexity and a small footprint. The main aim is to provide the protection of data at end-to-end device and expand the use of cryptography constrained devices. It is a cryptographic set of protocols or algorithm for implementing the constrained devices like sensors, RFID tags, smart cards, etc.

Such inestimable investigators obtain in study of [26] and [27] provides the development of lightweight cryptography as a result for in the asset of security used devices in IoT used application, as similarly as to stay aware of security of information for more approval.

7.5.3 Biometrics

Biometric is the measured statistical analysis of human beings' unique physical and behavioral attributes. This is mainly used for the identification and authentication of individuals. As we know, smart cities and homes provide interconnected IoT devices which access, transmit and store individual information. For providing the security in user device interface, biometric authentication is required, which can be performed using thumb impression, speech or face recognition. One of the protected approval plans provided in IoT used smart cities and smart homes is biometric affirmation. Based upon this scenario, we can include character issues in sagacious conditions in an IoT-used environment in two parts: (1) conspicuous verification, (2) approval. This can be invented in [28], inspect a couple of benefits of the biometric security system in different IoT-used scenarios.

7.5.4 Machine Learning

Machine Learning is the computer-based intelligence to learn without being programmed explicitly. In this it performs similar activity to human decision-making and behavior by providing them the capability to develop and learn their own program. For this, the huge amount of data provided as input to machine and different algorithms are used to build a machine learning model. The model given in [29], the makers analyze the benefits of using AI development to give security in WSNs. Furthermore, in [30] a unique model taking into account ML estimations was made to perceive different attacks in Wi-Fi associations. In addition, a few ML methods, as overseen that provide solo learning used [31] to perceive the presence of a botnet.

7.6 Conclusion

IoT-based smart cities and smart homes can further develop individuals' personal satisfaction and prosperity. IoT, as a trend-setting innovation executed in a smart city, smart home, smart vehicle and so on, can empower programming, wearable gadgets, and different items to convey and share data over the web.

In spite of this, the IoT framework and gadgets are powerless, shaky and unequipped for self-assurance and guard for an assortment of reasons, including restricted assets for IoT gadgets, improvement and organization issues, an absence of secure programming and equipment plan, and the variety of IoT assets.

Therefore, smart cities and smart homes have become powerless against a wide scope of safety dangers that put information security in danger of infringement by pernicious people. That makes security and protection a central issue that requires proficient and compelling arrangements that should guarantee the validation, accessibility, uprightness and secrecy of information communicated in the IoT organization.

As an outcome, it is vital to consider the security and protection dangers when planning and carrying out new smart frameworks. In this paper, we researched and examined security dangers in different layers of IoT design. Then, at that point we introduced a few arrangements and instruments for getting smart conditions dependent on IoT innovation.

References

1. Ikrissi, G., Mazri, T., 2020a. A Study of Smart Campus Environment and Its Security Attacks. *ISPRS - International Archives of the Photogrammetry,*

Remote Sensing and Spatial Information Sciences XLIV-4/W3-2020, 255–261. https://doi.org/10.5194/isprs-archives-XLIV-4-W3-2020-255- 2020.

2. Ikrissi, G., Mazri, T., 2020b. A Study of Smart Campus Environment and Its Security Attacks. *ISPRS - International Archives of the Photogrammetry, Remote Sensing and Spatial Information Sciences* XLIV-4/W3-2020, 255–261. https://doi.org/10.5194/isprs-archives-XLIV-4-W3-2020-255-2020.
3. H. Lin and N. W. Bergmann, "IoT privacy and security challenges for smart home environments," *Information*, vol. 7, no. 3, p. 44, 2016.
4. V. R. Kebande, N. M. Karie, and H. S. Venter, "Adding digital forensic readiness as a security component to the IoT domain," *Int. J. Adv. Sci., Eng. Inf. Technol.*, vol. 8, no. 1, pp. 1–11, 2018.
5. Al-Turjman, F., Zahmatkesh, H., Shahroze, R., 2019. An overview of security and privacy in smart cities' IoT communications. *Transactions on Emerging Telecommunications Technologies*. https://doi.org/10.1002/ett.3677.
6. Gubbi, R. Buyya, S. Marusic, and M. Palaniswami, "Internet of Things (IoT): A vision, architectural elements, and future directions," *Future Gener. Comput. Syst.*, vol. 29, pp. 1645–1660, 2013.
7. I. Batra, S. Verma, Kavita, and M. Alazab, "A lightweight IoT-based security framework for inventory automation using wireless sensor network," *Int. J. Commun. Syst.*, vol. 33, no. 4, p. e4228, Mar. 2020.
8. K. Djemame, D. Armstrong, M. Kiran, and M. Jiang, "A risk assessment framework and software toolkit for cloud service ecosystems," in *Proc. Int. Conf. Cloud Comput., GRIDs, Virtualization*, 2011, pp. 119–126.
9. W. M. Kang, S. Y. Moon, and J. H. Park, "An enhanced security framework for home appliances in smart home," *Hum.-Centric Comput. Inf. Sci.*, vol. 7, no. 1, Dec. 2017, Art. no. 6, doi: 10.1186/s13673-017-0087-4.
10. T. Denning, T. Kohno, and H. M. Levy, "Computer security and the modern home," *Commun. ACM*, vol. 56, no. 1, pp. 94–103, Jan. 2013, doi:10.1145/2398356.2398377.
11. A. Hinduja and M. Pandey, "An ANP-GRA-based evaluation model for security features of IoT systems," in *Intelligent Communication, Control and Devices* (Advances in Intelligent Systems and Computing), vol. 989, S. Choudhury, R. Mishra, R. Mishra, and A. Kumar, Eds. Singapore: Springer, 2020, doi: 10.1007/978-981-13-8618-3_26.
12. Taryudi, D. B. Adriano, and W. A. C. Budi, '"IoT-based integrated home security and monitoring system,"' *J. Phys., Conf. Ser.*, vol. 1140, no. 1, Dec. 2018, Art. no. 012006.
13. X. Li, X. Jin, Q. Wang, M. Cao, and X. Chen, "SCCAF: A secure and compliant continuous assessment framework in cloud-based IoT context," *Wireless Commun. Mobile Comput.*, vol. 2018, Oct. 2018, Art. no. 3078272, doi: 10.1155/2018/3078272.
14. Hassija, V., Chamola, V., Saxena, V., Jain, D., Goyal, P., Sikdar, B., 2019. A Survey on IoT Security: Application Areas, Security Threats, and Solution

Architectures. *IEEE Access* 7, 82721–82743. https://doi.org/10.1109/ACCESS.2019.2924045.

15. Khan, M.A., Salah, K., 2018. IoT security: Review, blockchain solutions, and open challenges. *Future Generation Computer Systems* 82, 395–411. https://doi.org/10.1016/j.future.2017.11.022.
16. Ahemd, M.M., Shah, M.A., Wahid, A., 2017. IoT security: layered approach for attacks defenses, in: *2017 International Conference on Communication Technologies (ComTech)*. Presented at the 2017 International Conference on Communication Technologies (ComTech), pp. 104–110. https://doi.org/10.1109/COMTECH.2017.8065757.
17. Al-Turjman, F., Zahmatkesh, H., Shahroze, R., 2019. An overview of security and privacy in smart cities' IoT communications. *Transactions on Emerging Telecommunications Technologies*. https://doi.org/10.1002/ett.3677.
18. Vishwakarma, R., Jain, A.K., 2020. A survey of DDoS attacking techniques and defence mechanisms in the IoT network. *Telecommun Syst* 73, 3–25. https://doi.org/10.1007/s11235-019- 00599-z.
19. Kamble, A., Malemath, V.S., Patil, D., 2017. Security attacks and secure routing protocols in RPL-based Internet of Things: Survey, in: *2017 International Conference on Emerging Trends Innovation in ICT (ICEI)*. Presented at the 2017 International Conference on Emerging Trends Innovation in ICT (ICEI), pp. 33–39. https://doi.org/10.1109/ETIICT.2017.7977006.
20. Hafeez, I., Antikainen, M., Tarkoma, S., 2019. Protecting IoT environments against Traffic Analysis Attacks with Traffic Morphing, in: *2019 IEEE International Conference on Pervasive Computing and Communications Workshops (PerCom Workshops)*. Presented at the 2019 IEEE International Conference on Pervasive Computing and Communications Workshops (PerCom Workshops), pp. 196–201. https://doi.org/10.1109/PERCOMW.2019.8730787.
21. Kumar, R.P., Smys, S., 2018. A novel report on architecture, protocols and applications in Internet of Things (IoT), in: *2018 2nd International Conference on Inventive Systems and Control (ICISC)*. Presented at the 2018 2nd International Conference on Inventive Systems and Control (ICISC), pp. 1156–1161. https://doi.org/10.1109/ICISC.2018.8398986.
22. Rizvi, S., Kurtz, A., Pfeffer, J., Rizvi, M., 2018. Securing the Internet of Things (IoT): A Security Taxonomy for IoT, in: *2018 17th IEEE International Conference on Trust, Security and Privacy in Computing and Communications/12th IEEE International Conference on Big Data Science and Engineering (TrustCom/BigDataSE)*. Presented at the 2018 17th IEEE International Conference on Trust, Security and Privacy in Computing and Communications/12th IEEE International Conference on Big Data Science and Engineering (TrustCom/BigDataSE), pp. 163–168. https://doi.org/10.1109/TrustCom/BigDataSE.2018.00034.

23. Abdulghani, H.A., Nijdam, N.A., Collen, A., Konstantas, D., 2019. A Study on Security and Privacy Guidelines, Countermeasures, Threats: IoT Data at Rest Perspective. *Symmetry* 11, 774. https://doi.org/10.3390/sym11060774.
24. Abdul-Ghani, H.A., Konstantas, D., Mahyoub, M., 2018. A Comprehensive IoT Attacks Survey based on a Building-blocked Reference Model. *International Journal of Advanced Computer Science and Applications (IJACSA)* 9. https://doi.org/10.14569/IJACSA.2018.090349.
25. Swamy, S.N., Jadhav, D., Kulkarni, N., 2017. Security threats in the application layer in IOT applications, in: *2017 International Conference on I-SMAC (IoT in Social, Mobile, Analytics and Cloud) (I-SMAC)*. Presented at the 2017 International Conference on I-SMAC (IoT in Social, Mobile, Analytics and Cloud) (ISMAC), pp. 477–480. https://doi.org/10.1109/ISMAC.2017.8058395.
26. Bhardwaj, I., Kumar, A., Bansal, M., 2017. A review on lightweight cryptography algorithms for data security and authentication in IoTs, in: *2017 4th International Conference on Signal Processing, Computing and Control (ISPCC)*. Presented at the 2017 4th International Conference on Signal Processing, Computing and Control (ISPCC), pp. 504–509. https://doi.org/10.1109/ISPCC.2017.8269731.
27. Dhanda, S.S., Singh, B., Jindal, P., 2020. Lightweight Cryptography: A Solution to Secure IoT. Wireless Pers Commun 112, 1947–1980. https://doi.org/10.1007/s11277-020-07134-3
28. Elrawy, M.F., Awad, A.I., Hamed, H.F.A., 2018. Intrusion detection systems for IoT-based smart city and smart homes: a survey. *Journal of Cloud Computing* 7, 21. https://doi.org/10.1186/s13677-018-0123-6.
29. Obaidat, M.S., Rana, S.P., Maitra, T., Giri, D., Dutta, S., 2019. Biometric Security and Internet of Things (IoT), in: Obaidat, M.S., Traore, I., Woungang, I. (Eds.), *Biometric-Based Physical and Cybersecurity Systems*. Springer International Publishing, Cham, pp. 477–509. https://doi.org/10.1007/978-3-319-98734-7_19.
30. Aminanto, M.E., Choi, R., Tanuwidjaja, H.C., Yoo, P.D., Kim, K., 2018. Deep Abstraction and Weighted Feature Selection for Wi-Fi Impersonation Detection. *IEEE Transactions on Information Forensics and Security* 13, 621–636. https://doi.org/10.1109/TIFS.2017.2762828.
31. Alsheikh, M.A., Lin, S., Niyato, D., Tan, H.-P., 2014. Machine Learning in Wireless Sensor Networks: Algorithms, Strategies, and Applications. *IEEE Communications Surveys Tutorials* 16, 1996–2018. https://doi.org/10.1109/COMST.2014.2320099.
32. Miller, S., Busby-Earle, C., 2016. The role of machine learning in botnet detection, in: *2016 11th International Conference for Internet Technology and Secured Transactions (ICITST)*. Presented at the 2016 11th International Conference for Internet Technology and Secured Transactions (ICITST), pp. 359–364. https://doi.org/10.1109/ICITST.2016.7856730.

8

Traffic Management for Smart City Using Deep Learning

Puja Gupta* **and Upendra Singh**

Department of Information Technology, S.G.S.I.T.S. Indore, India

Abstract

Recent decades have seen an increase in the population density of cities, and today's megacities are densely inhabited locations with substantial land use encompassing residential, transit, sanitation and utilities as well as communication infrastructure. In the future, new types of road users will emerge as a consequence of new technology developments. As cities grow in population and highways become more congested, governmental agencies such as the Transportation Department and the National Highway Administration are under increasing pressure to improve their management services through the introduction of more efficient technology. In order to preserve life and establish long-term, expense management methods, the objective is to forecast issues that have never previously been faced. Self-driving vehicles will soon be legal in highly populated major cities where streets will be shared by pedestrians, bicycles, autos, and trucks, further complicating the situation. Adaptations to road and signal widths and timing will be required on a regular basis. Human involvement is still required to count and categorize turning vehicles and pedestrians at intersections, even with the use of traffic monitoring technology. Turning-vehicle counts at traffic junctions may be resolved using our method, which is less invasive, requires no road excavating, and is less costly to implement. Cameras installed across cities, as well as any other camera, including those on mobile phones, may be used to capture video in real time or after they have been captured. Neural Networks and Deep Learning, as well as machine vision and a range of techniques and algorithms, are all used in our system to identify objects. An artificial convolutional neural network will be used to analyze moving objects in both still images and video recordings, allowing us to follow their motion and calculate their velocity as well as direction.

**Corresponding author*: pooja1porwal@gmail.com

Romil Rawat, A. Mary Sowjanya, Syed Imran Patel, Varshali Jaiswal, Imran Khan and Allam Balaram (eds.) Autonomous Vehicles Volume 1: Using Machine Intelligence, (149–160) © 2023 Scrivener Publishing LLC

Keywords: Computer vision, deep learning, neural networks, convolutional neural networks, detection, tracking

8.1 Introduction

Smart cities are defined as Indian municipalities that employ information and communications technologies (ICT) to increase efficiency and productivity, share data with the public at large, and improve government services and the quality of life of its residents [1]. Data gathered from traffic sensors is crucial to the knowledge of and improvement in safety. The public road system is monitored by a variety of devices, including magnetic loops, pressure tubes, radar guns, microwave sensors, and cameras, all of which are used by modern governments for the purpose of gathering traffic data. In order to fully understand and solve the safety issue, data on street and highway accidents must be analyzed in every state and municipality [3].

The adoption of traffic surveillance systems will help all sectors, including governments, industry, and individuals, since they will be able to make better judgments based on the data gathered. For a smart city to be really successful, it has to concentrate on the well-being of its residents and visitors, as well as attracting businesses and talent, and ensuring that the environment is protected [2]. Economic benefits may be gained by making better decisions about where to invest money on safety enhancement projects thanks to accurate traffic data collection. The first stage in creating smart cities is gathering accurate data and then analyzing that data to come up with long-lasting solutions. Transportation data collection may be challenging because of the human participation and volume of data that must be analyzed, especially when video images or real traffic photography are involved.

Deep-learning, which is based on feature learning, is a method of using ml algorithms to uncover highly representative characteristics from input data that is based on feature learning (for example, an image). For object recognition, convolutional neural networks (CNNs) are the deep learning technology most often utilized [4–6]. Chen *et al.* [7] proposed a vehicle detection method using template matching and deep CNNs. In order to detect autos at different scales, the deep network uses convolution layers that are repeated in several sizes. Auto detection algorithm training on a GPU requires large computational resources. With CNN and sliding windows, determining the placement of stuff is a challenge [8]. A number of other approaches have been proposed to address this issue, including Sppnet [9], R-CNN [10], and Fast R-CNN [11]. On the other hand, the creation of region suggestions requires an exorbitant amount of computation time. R-CNN and Faster R-CNN [12] were created by Ren *et al.* with the purpose of providing the most accurate item detection accuracy while also assuring real-time detection speed. Traffic data and information study have shown that they are limited in their capacity to capture traffic data at junctions in an impartial and acceptable manner. Non-linear crossings may be difficult to monitor because of a lack of reliable traffic monitoring technologies like magnetic loops and pressure tubes as well as radar guns and microwave sensors. A lack of separate counting systems was found at orthogonal crossings, which we felt was strange. A few studies or demonstrations of traffic data collection for orthogonal crossings that employ a left-turn, right-turn, and drive-throughs counting method have been conducted. We can learn a lot about the need for improved traffic monitoring systems from a number of attempts to make the lives of those who collect traffic data a bit simpler. Individual components of the system that we've developed and tested so far have been demonstrated to be extremely promising in our research and early tests in developing such an intensive monitoring system.

An investigation of the gathering of valuable information from obtained data is the goal of this paper. Data traffic statistics may be collected using a wide range of technologies, but we must realize that even with the most modern technology, people play an important role and such operations are time demanding. We want to decrease or eliminate the need for human intervention in the computing process by implementing a vision-based and deep learning-based autonomous surveillance system. Any moving item may be tracked and classified using a system like this, which takes into consideration the object's position as well as its senses and speed. The human component must be eliminated wherever feasible in order to acquire accurate and trustworthy data.

8.2 Literature Review

For car recognition in digital pictures, the Viola-Jones (VJ) [13] and Discriminatively Trained Part Based Models (DPM) [14] object detection approaches have been applied. For vehicle identification, these methods are less vulnerable than others in the literature to image noise and more complex situations. The majority of such approaches, on the other hand, are extremely sensitive towards the rotation of the item. Some technologies like virtual reality (VJ), on the other hand, have a high sensitivity to changes in light.

There are a number of machine learning-based detection methods that have been published in the literature. In virtual photographs, car identification is more difficult than other item identifying facts because of the very high quality of the images. The SVM classifier was employed to categorize the recovered objects of interest into two groups based on their characteristics: "Vehicle" and "no-Vehicle," based on Moranduzzo and Melagni's [15] suggestion of a scale-invariant features transformation (SIFT) to identify the keypoints of automobiles. Each SIFT point associated with a specific vehicle is then integrated to create an overall picture of that vehicle. An object recognition method for UAV photographs combining higher-order gradient and Gaussian process regression has been developed by Moranduzzo *et al.* [16], who applied it to the challenge of identifying cars in digital images taken inside cities. As well as histograms of gradient and the SVM classifier, Xu *et al.* [17] employed a combination of the two to recognize automobiles in drone-shot street video.

Recently, a number of techniques for vehicle recognition have been suggested using deep learning. As a result of this research, Perez *et al.* [18] created a hitherto unknown classical object recognition system based just on the sliding window technique. As a result of this method, it is very time consuming to identify multi-scale objects utilizing the sliding window technique. Chen *et al.* [5] proposed a technique for recognizing cars in satellite images using CNN-based deep learning. A feature extractor developed by Girshick *et al.* [10] was used to extract a subset of areas from an image, which they referred to as region recommendations, so that we may avoid the difficulties associated with retrieving a large range of areas using conventional CNN models. Ammour *et al.* [19] devised an algorithm for car identification that included three stages: generating candidate regions, extracting features, and classifying. To begin, images were segmented using the mean-shift approach [20], and then area information was extracted using fine-tuned VGG16 models [21]. It is then used to categorize the areas. In order to come up with a list of potential candidate places, the procedure is lengthy. Furthermore, each of the three processes requires the training of several models, which adds to the model's complexity. Convolutional support vector machine (CSVM) networks were developed by Bazi and colleagues [22] to identify vehicles in unmanned aerial vehicle (UAV) data. There are several convolution and reduction layers that alternate with each other and end with the linear SVM classifier in CSVM. Ren *et al.* [12] offered Faster R-CNN as a promising technique for the time-consuming operation of region proposal development with R-CNN approaches to ease this difficulty. The performance of a Faster R-CNN technique in terms including both recognition rate and detection speed was encouraging, which led to such an investigation. Consequently, A Faster R-CNN approach is used to choose a large number of autos from digital pictures.

Faster R-CNN-based vehicle identification

The Regional Proposal Network (RPN) and the Fast R-CNN detector make up the Faster R-CNN (FR-CNN), as shown in Figure 8.1. When producing region proposals, the RPN uses a neural network to generate region proposals that are available in a broad variety of shapes and sizes and dimensions. These ideas will be sent to the next group. Depending on the proposal, proposals might contain or exclude target objects. The suggestions are fine-tuned and improved using this technology.

The FR-CNN only travels through all the CNN once for the whole input images, and so it updates the objects recommendations that were previously generated.

I. Regional Proposal Networks: This network uses a digital picture for input as well outputs with a series of frames, each carrying an item better prediction from each of the image's classes. VGG16, ResNet50, and GoogLeNet convolutional backbends are used in this application. The RPN creates suggestions for rectangular areas at each location using sliding windows and a convolutional feature map created by the final neural layer. Non-maximum suppressing (NMS) has been used to integrate ideas that have a significant crossing percentage because numerous proposals have a lot of overlap with one another (IoU). Next, the remaining ideas are

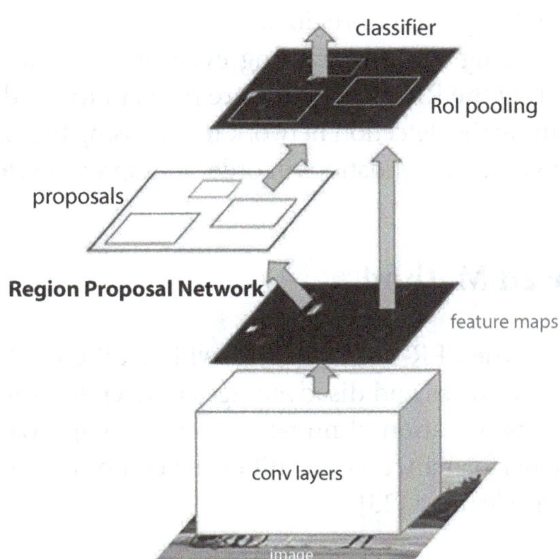

Figure 8.1 Detection of objects using a faster R-CNN.

ranked in order of their object likelihood scores, only the top-ranked ideas can be used for detection, with the remaining proposals being discarded.

II. Fast R-CNN detectors: Many areas of interest may be fed into the FR-CNN (RoIs). The RoI Pooling of Convolution Layers collects a fixed-length feature representation for every area of interest, which is then used in the next step (RoI). Layered architecture feeds every distinguishing feature into a series of connected layers. With the use of a softmax as well as bounding-box regressor layer, softmax probability are produced that measure across K objects categories, as well as bounding-box values that correlate to those softmax probabilities.

III. Fast R-CNN Training: When RPNs are trained, each suggestion is issued a K+1 classifier that defines if the suggestion is indeed an item from one of the K automobile kinds. This class label is used to train RPNs (background). The fact that RPN as well as FR-CNN use the same convolution layer means that they'll be learned with each other to build a single system, which is advantageous. Steps 1 through 4 of the Faster R-CNN model [12] training process comprise the following:
   - The RPN network should be trained in the same manner as described previously.
   - Using the region suggestions created by that of the RPN trained in the first phase, train the detection network using the proposed regions.
   - Initializing all RPN training even by detector, but only training the RPN layers that are relevant to the detection.
   - Training the detection network model using the RPN's new region recommendations in order to improve performance.

8.3 Proposed Method

As a consequence, the FPR-FNR trade-off will be different for each model because of the advantages and disadvantages of deep learning and classifications. Using a combination of numerous learning approaches, it is possible to improve performance as a result of the combination of the outputs from different base learner [23].

- Paradigm of transfer of learning: For car identification, this study used transfer learning since the Faster R-CNN training technique requires a broad variety of datasets as well as considerable computational resources [24]. Transfer-based learning is used to build a more accurate system in a shorter period of time. This is due to the fact that transfer-based learning does not begin from base, but rather with systems that have already handled a range of problems. The proposed model utilizes the present Faster R-CNN model's convolutional basis as its characteristics extract and adds again new classifier on top of upgraded model for improved classification accuracy. While the output of pre-trained models like VGG16 has 1,000 different class labels, our newly built model only contains five distinct object classes (vehicle, truck, lorry, buses, and trailer). Sample data with tagged photographs may be used to train the new classifier's variables and then used to train the classifier itself. It was possible to recognize cars utilizing a feature extractor trained on an R-CNN and an output provided by a newly developed classifier at the top-level using a single transfer learning framework.

In order to implement a transfer-learning paradigm, there are three primary steps that must be taken.

- Several models that have been pre-trained on the ImageNet are chosen (VGG16, ResNet50).
- When used as a feature extractor, the convolutional base from these models is extracted one at a time, and it is then combined with the newly defined classification layer to create a fresh Faster R-CNN models.
- This updated classification is performed with the dataset, which is a samples picture dataset, and then evaluated using the newly created dataset. Thus, the weights of the convolutional layer base remain constant, and even the variables like an extra classifiers layers were learnt during this phase of the learning process.

8.4 Experimental Evaluation

8.4.1 Hardware and Software Configuration

Tests of suggested extended MRCNN approaches for detecting social distance are carried out using the following set-up. An Intel Core i7-1065G7

CPU running at 3.9 GHz and a 15.6-inch high-definition LED touch screen are just some of the features that come with the Python interface. A 1 terabyte hard drive. Intel Iris has 8GB of RAM, as well as graphics and twin headphones. HD camera from Hewlett-Packard Company. All of these features are included in the Realtek RTL8821CE Python programming was finished on Windows 10's 64-bit edition of the OS. Pandas, Numpy, Scikit-Learn, Matplotlib, Scikit-Learn, Plotly, Tensor Flow, Keras, and Seaborn were all utilized throughout the implementation process.

8.4.2 About Dataset

Indore bypass road and AB road near Indore are two new datasets that we collected from cameras watching traffic in the city of Indore, India. We sample a frame every two seconds from several video streams in order to guarantee that the data are not connected in time.

8.4.3 Implementation

After developing code on a video-based dataset, it is necessary to share implementation findings and screenshots during this step. Identification and categorization of road traffic data, as well as the collection of vehicle number plates, is beneficial to RTO. Figure 8.2 shows the image scenario to identify all vehicle and scan traffic on road and Figure 8.3 represents models to identify cars (vehicles) on road.

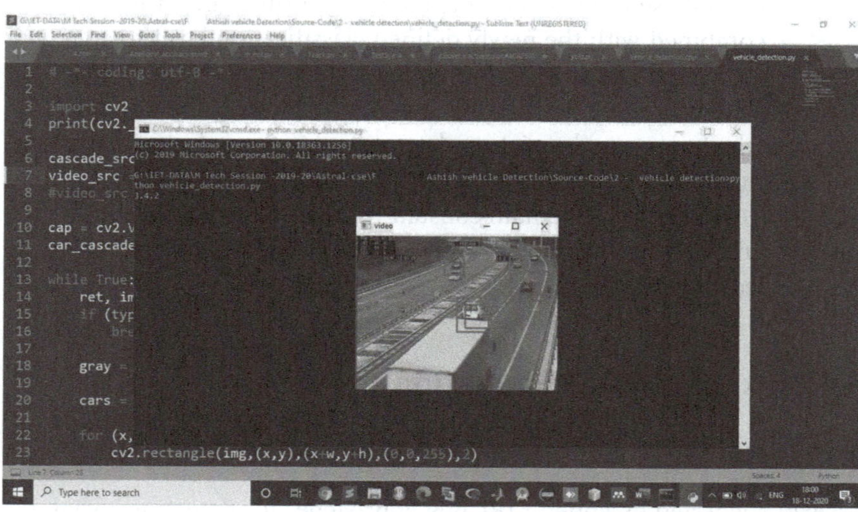

Figure 8.2 Identify all vehicle and scan traffic on road.

Figure 8.3 Identify cars (vehicles) on road.

8.4.4 Result

In this part, we will compute the root mean square error on the proposed model with regard to the number of hidden layers that have been included. As shown in Figure 8.4, we must raise the number of concealed layers from one to thirty.

Root Mean Square Error (RMSE) =

$$\text{RMSE} = \sqrt{\sum_{i=1}^{n} \frac{\left(\hat{y}_i - y_i\right)^2}{n} \frac{1}{2}}$$

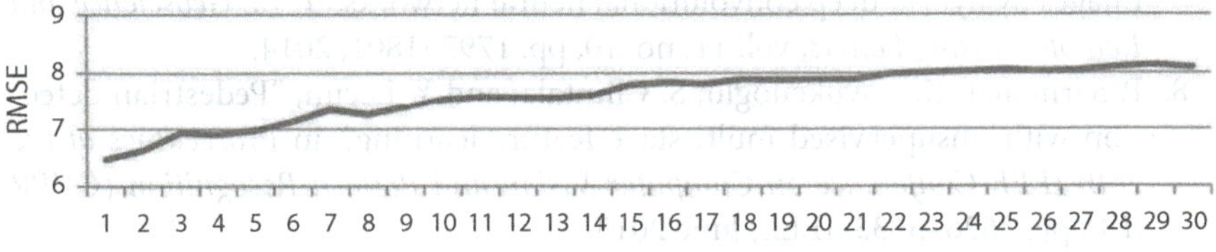

Figure 8.4 RMSE value in respect of number of hidden layers.

8.5 Conclusion

A deep learning-based system for vehicle recognition from digital pictures has been created and shown in this work. Region proposals were derived through RPN, and afterwards CNN is being used to categorize them into vehicle or no object classes depending on where they are placed in the region. This is done in accordance with Faster R-CNN object recognition technique (background). Three CNNs have been used to extract features in the transfer-learning model, which has been combined with five other classifiers to get the final result. A recent publishing image dataset (AU-AIR) test showed that the model beat earlier techniques in terms of precision, accuracy and recall. This was confirmed by simulations on a similar dataset. A majority vote ensemble aggregates the various basis learning, resulting in more efficient learning. For future machine learning research, a weighted average ensemble, which is optimized via heuristic algorithms (such as the use of a genetic algorithm), may be used.

References

1. Smart city definition https://internetofthingsagenda.techtarget.com/definition/smart-city
2. Smart cities – Deloitte, https://www2.deloitte.com/insights/us/en/focus/smartcity/overview.html Accessed 2018-11-11 22:58:39.
3. National Safety Conference – 1924 Washington, D.C.
4. Huang, G. L., He, J., Xu, Z., & Huang, G. (2020). A combination model based on transfer learning for waste classification. *Concurrency and Computation: Practice and Experience*, e5751.
5. El Mostafa Hambi, F. B. A New Online Plagiarism Detection System based on Deep Learning. *International Journal of Advanced Computer Sciences and Applications*, 11(9), 470-478.
6. Shustanov, A., & Yakimov, P. (2017). CNN design for realtime traffic sign recognition. *Procedia Engineering*, 201, 718-725.
7. X. Chen, S. Xiang, C.-L. Liu, and C.-H. Pan, "Vehicle detection in satellite images by hybrid deep convolutional neural networks," *IEEE Geoscience and Remote Sensing Letters*, vol. 11, no. 10, pp. 1797–1801, 2014.
8. P. Sermanet, K. Kavukcuoglu, S. Chintala, and Y. Lecun, "Pedestrian detection with unsupervised multi-stage feature learning," in *Proceedings of the 26th IEEE Conference on Computer Vision and Pattern Recognition (CVPR '13)*, pp. 3626–3633, IEEE, June 2013.
9. K. He, X. Zhang, S. Ren, and J. Sun, "Spatial pyramid pooling in deep convolutional networks for visual recognition," *IEEE Transactions on Pattern Analysis and Machine Intelligence*, vol. 37, no. 9, pp. 1904–1916, 2015.

10. R. Girshick, J. Donahue, T. Darrell, and J. Malik, "Rich feature hierarchies for accurate object detection and semantic segmentation," in *Proceedings of the 27th IEEE Conference on Computer Vision and Pattern Recognition*, pp. 580–587, Columbus, OH, USA, 2014.
11. R. Girshick, "Fast R-CNN," in *Proceedings of the 15th IEEE International Conference on Computer Vision (ICCV '15)*, pp. 1440–1448, December 2015.
12. Ren, S., He, K., Girshick, R., & Sun, J. (2016). Faster RCNN: towards real-time object detection with region proposal networks. IEEE transactions on pattern analysis and machine intelligence, 39(6), 1137-1149.
13. P. Viola and M. Jones, "Rapid object detection using a boosted cascade of simple features," in *Proceedings of the IEEE Computer Society Conference on Computer Vision and Pattern Recognition*, pp. 511–518, December 2001.
14. P. F. Felzenszwalb, R. B. Girshick, D. McAllester, and D. Ramanan, "Object detection with discriminatively trained partbased models," *IEEE Transactions on Pattern Analysis and Machine Intelligence*, vol. 32, no. 9, pp. 1627–1645, 2010.
15. Zhao, T.; Nevatia, R. Car detection in low resolution aerial image. In *Proceedings of the Eighth IEEE International Conference on Computer Vision (ICCV 2001)*, Washington, WA, USA, 7–14 July 2001.
16. Moranduzzo, T.; Melgani, F. Automatic Car Counting Method for Unmanned Aerial Vehicle Images. *IEEE Trans. Geosci. Remote Sens.* 2014, 52, 1635–1647.
17. Xu, Y.; Yu, G.; Wang, Y.; Wu, X.; Ma, Y. A Hybrid Vehicle Detection Method Based on Viola-Jones and HOG + SVM from UAV Images. *Sensors* 2016, 16, 1325.
18. A. Perez, P. Chamoso, V. Parra, and A. J. Sanchez, "Ground vehicle detection through aerial images taken by a UAV," in *Proceedings of the 17th International Conference on Information Fusion, (FUSION '14)*, July 2014.
19. N. Ammour, H. Alhichri, Y. Bazi, B. Benjdira, N. Alajlan, and M. Zuair, "Deep learning approach for car detection in UAV imagery," *Remote Sensing*, vol. 9, no. 4, p. 312, 2017.
20. D. Comaniciu and P. Meer, "Mean shift: a robust approach toward feature space analysis," *IEEE Transactions on Pattern Analysis and Machine Intelligence*, vol. 24, no. 5, pp. 603–619, 2002.
21. K. Simonyan and A. Zisserman, "Very deep convolutional networks for large-scale image recognition," *Computer Science*, 2014.
22. Bazi, Y., & Melgani, F. (2018). Convolutional SVM networks for object detection in UAV imagery. *IEEE transactions on geoscience and remote* sensing, 56(6), 3107-3118.
23. Shair, E. F., Jamaluddin, N. A., & Abdullah, A. R. (2020). Finger Movement Discrimination of EMG Signals towards Improved Prosthetic Control using TFD. *International Journal of Advanced Computer Sciences and Applications*, 11(9), 244-251.
24. Bozcan, I., & Kayacan, E. (2020). AU-AIR: A Multi-modal Unmanned Aerial Vehicle Dataset for Low Altitude Traffic Surveillance. arXiv preprint arXiv:2001.11737.

9

Cyber Security and Threat Analysis in Autonomous Vehicles

Siddhant Dash* and Chandrashekhar Azad

Department of Computer Science Engineering, NIT Jamshedpur, India

Abstract

Autonomous vehicles are driverless vehicles that can operate themselves using complicated driverless and self-sustained systems. They are able to operate by sensing their surrounding and performing calculations without human intervention. With global IoT systems under constant attack, autonomous systems have also seen a drastic rise in security concerns. There can be lots of attack vectors that can be targeted in a fully automated vehicle. Some of the attack surface are Smart keys, Vehicle data servers, Software updates, ECUs, wireless communication systems. All these can be exploited and can potentially lead to further greater losses. Attacks can range from simple denial of access or locking out of systems to vehicle thefts, accidents, and data thefts. With the hackers gaining remote access to an operational vehicle there can be some serious threat to life. This book chapter examines all the details regarding security and cyber threats to autonomous vehicles and their potential impact. Different systematic approaches like system security auditing, gap analysis, security assessment, and data trust centres can help to minimize the attacks and protect autonomous systems. The severity of this situation needs to be tackled immediately and swiftly by car manufacturers.

Keywords: Autonomous vehicle, cyber security, threat analysis, security systems, wireless attacks, internal network attacks, external network attacks

**Corresponding author*: 2021pgcais03@nitjsr.ac.in

Romil Rawat, A. Mary Sowjanya, Syed Imran Patel, Varshali Jaiswal, Imran Khan and Allam Balaram (eds.) Autonomous Vehicles Volume 1: Using Machine Intelligence, (161–180) © 2023 Scrivener Publishing LLC

9.1 Introduction

"Self-driving cars are the natural extension of active safety and obviously something we think we should do", said Elon Musk. These systems are the technology of future as they propel our automotive industry to new heights. The global autonomous vehicle market is set to grow at an exponential pace. With the major car manufacturing industry understanding the trend, they will soon compete for dominance in this sector. An estimated figure suggests that the global AV industry is set to grow up to $7 trillion by 2050. This figure is enough to provide the scale and speed with which it will go.

With all these potential benefits come serious cyber threats and security loopholes. The connectivity and enhanced experience introduce complex and even complicated software issues. The difficulty in finding and patching these would grow significantly higher. Security premises of these new techs are often ignored or not taken seriously. The software running these codes could be hacked by using different attack vectors for some known vulnerability. Recently, more than a dozen vulnerabilities in a Mercedes-Benz E-Class car were discovered by security researchers from the Sky-Go Team, Qihoo 360's automotive hacking squad, allowing them to remotely open the doors and start the engine [2]. This shows what the adversaries are capable of and their potential.

These attacks lead to the parent company researching and patching the software against these vulnerabilities. Another potential way to introduce malware in an AV system can be external USB drives. Potential attack possibilities are endless.

We discuss major security attacks and threats to autonomous vehicle systems, challenges faced in this sector and related research methodologies and proposals. The major works and contributions to this sector will provide us with a clear overview about the current attack scenario, our defence mechanisms and strategies and future work for its improvement.

9.2 Autonomous Vehicles

Autonomous vehicles (AVs) or self-driving cars are independent vehicle systems capable of movement with minimal human intervention. An AV senses its surroundings using different sensors like Radar, Lidar, Sonar and

GPS. Its control system gets input from the surroundings and accordingly maps its path for safe travel avoiding obstacles and people. Advanced data science and machine learning are core technology for training the model systems for real-world scenarios.

Development in processing power and computational ability made high bandwidth and real-time data transfer to and from online command centres. Different projects of autonomous vehicle development are in different stages. Society of Automotive Engineers classified these developments and relative autonomy into 6 levels with level 0 being zero automation and level 5 as steering wheel optional.

9.2.1 Autonomous vs. Automated

Autonomous and Automated vehicles are quite similar in their autonomy but vastly different in their approach. Automated vehicles are preprogrammed in certain ways to follow a destined path and in a specific manner, whereas autonomous vehicles are independent in their mobility aspect with very little human intervention. After setting initial and final destination, routes and everything in between are handled by the vehicular system itself.

9.2.2 Significance of Autonomous Vehicles

The main motive for autonomous vehicle development is people security. Approximately 1.3 million people die each year due to road or vehicular accidents and around 20-50 million suffer non-fatal injuries due to the same. Government data suggests that 94% of accidents are caused by a human entity, either driver's fault or another external human factor. A higher degree of autonomy can ensure no erroneous human behaviour like rash driving, intoxicated, driving under other influence, speeding, impaired driving. Reduction in such cases can significantly improve communal life and safety.

Autonomous vehicles will not only reduce the number of accidents on the road but help in building smart cities as well. With the onset of IoT and smart connectivity, everyday objects are getting interconnected and inter-operable. Autonomous vehicles integrated with smart city systems would reduce traffic and other major issues. Autonomous Vehicular communication is not a thing of the past now as new studies have shown that local vehicle communication can significantly improve traffic condition of that area. Wenbo *et al.*, 2019 [9] introduced a coordination protocol for autonomous vehicles to coordinate their driving behaviour. The coordination would enable nearby vehicles to cooperate in a local group and agree upon driving plans for improving traffic efficiency.

Autonomous Vehicles will make the life of challenged people/people with special needs more accessible and fulfilling. People with injuries or disability tend to need a caretaker for daily chores and driving. With autonomous vehicles, such people will have an independent and fulfilling life. Senior citizens would enjoy the service as well.

With all being said, no projects are ever initiated without monetary benefits in mind. Autonomous vehicles will save a ton of money in the long term as they will improve efficiency of travel. The coordinated traffic system will halt less and have less congestions, thereby having low fuel consumption. Along with monetary, it incorporates environmental benefits. Low fuel consumption can lead to cleaner and greener cities.

9.2.3 Challenges in Autonomous Vehicles

Going with the current trend is easy but understanding the motive and implementing it on its own terms is hard. Autonomous vehicle technology may bring lots to the table for different companies and great future prosperity, but everything is not all roses. There are enormous challenges and obstacles to be tackled for developing a truly autonomous vehicle system.

Technological sustainability seems to be a main barrier for new entrants in this field. Companies need huge monetary resources [8], top-notch research facility backed by great employees for developing systems which can come close to level 5 of autonomy. This seems like the tech might end up being produced by a few, leading to monopoly in the worst case.

The other side of technological advancement lies in the risk of cyber attacks and online threats. Autonomous vehicles are no exception to these. All conventional attacks used on internet can be modified for attacking these systems. AV systems are at much higher risk due to their nature of operation and the complexity associated with this project. Security researchers have shown that it is possible to hack and compromise the autonomous system even remotely [2, 8]. Malware is still to be developed according to these systems, but it won't be too late when cyber criminals do catch up with this tech. Many security and academic researchers have pointed out enormous security problems and attack vectors associated with these autonomous vehicle systems.

Another interesting angle of challenge autonomous vehicles face is that of legal problems [10]. Every country tends to have their own road and traffic standards, laws and regulations. Autonomous vehicle systems need to cater to the needs of global customers. Any model trained on US roads will struggle to understand the regulations of Indian roads. The specificity of different laws and regulations need to be considered in advance.

Last but not least, of all the challenges and problems in developing autonomous systems, moral and ethical concerns cannot be ruled out. Autonomous systems are pieces of software running on codes; they have no moral obligations or ethical dilemma. Consider the following case: if an autonomous vehicle is travelling at high speed on a highway, suddenly someone unknowingly crosses the road and comes in front of the car. The deacceleration is not enough to stop without hitting something. Would the car steer away to avoid immediate collision and risk the driver's life or would it run into the person crossing? Such kinds of extreme cases and accidents need to be taken care of and handled with proper conscience.

9.2.4 Future Aspects

Even with all the challenges in the development of autonomous vehicles, all hope is not lost. Entrepreneurs, scientists and environmentalists are very enthusiastic about this tech and the way it will change our world. Ordinary people will benefit the most. These systems would lead to more interconnectivity in everyone's life. Everything would be interconnected by IoT and similar systems. With more systems being able to communicate, manual intervention would be reduced.

9.3 Related Works

Researchers from all around the globe are actively engaged in security investigation and research in autonomous vehicle sectors. The companies that build these systems and install them in vehicles have in-house security monitoring and management teams for following security guidelines and standards out of the box. Related papers have been published cataloguing and categorising the online and offline security threats to AV systems.

Brooks *et al.* (2009) [8] suggested a comprehensive study mapping all attacks, events and threats into various sections and providing a detailed flowchart diagram of it. The major thing highlighted in that paper is the inclusion of all possible events and their correlation with each other, ultimately how they affect the autonomous system's security. A takeover of the vehicle can affect the business integration platform as well. With the attacker having client-side role can leverage other server-side functionalities like tele diagnostics, remote system update and access to other third-party software which can prove fatal in any sense.

Yağdereli *et al.* (2015) [4] discussed the various levels of autonomy and their relative significance in security perspective. Different levels of vehicular autonomy attract different security problems and complications. There will always be a tradeoff between the degree of autonomy and security. More autonomous systems will have greater cyber threat risks, and so forth.

Sheehan *et al.* (2019) [7] provided a cyber-risk classification framework based on Bayesian network that utilizes CVSS software vulnerability risk-scoring framework. Their model can be used by regulatory bodies, manufacturers, suppliers and insurers for proactively assessing the cyber-risk potential of new and existing autonomous vehicle technology.

Palanca *et al.* (2017) [5] discussed denial-of-service attacks against automotive network. The attack proposed by them uses transmission of incomplete frames which would remain undetected from major security firewalls and frame-level analysis. The attack worked around CAN protocol weakness and showed the barrier against attacks are too low.

All these studies and works have shown the current scenario and depth of our security infrastructure. While some attacks and their vectors are relatively easy and straightforward, some need detailed and thorough analysis to discover. Major countries have been publishing important works and the amount of works are steadily growing. Industrial research and development have risen owing to the new attack methods unearthed daily like ransomware, phishing and other wireless attacks.

9.4 Security Problems in Autonomous Vehicles

As mentioned in the earlier section regarding the challenges faced by autonomous vehicles, security concerns stand at the top. The security attacks and threats are imminent for these systems and their consequences are staggering. While analysing security threats, it becomes essential to understand the security landscape of these vehicular systems.

Security landscape defines the overall components and how they affect security. It includes the environment in which they operate, other entities they interact with and their own systems. Based on these premises, different attack surfaces would be defined that facilitate hackers or a malicious person to compromise the security. The attack surface of a system is a collection of all different attack nodes, where the malicious actor can inject or extract sensitive information to and from the system thereby compromising its confidentiality and integrity. For example, using a compromised radar which is part of the vehicular system and acts like attack surface, a remote sensor attack can be launched. So, attack surfaces play an important role in both the offensive and defensive side of security. It is the main attack point for the bad actors as well as a monitoring point for security researchers to tighten the security. Different entities interact with each other at surfaces in the security landscape. Different attacks on autonomous vehicles are represented in Figure 9.1.

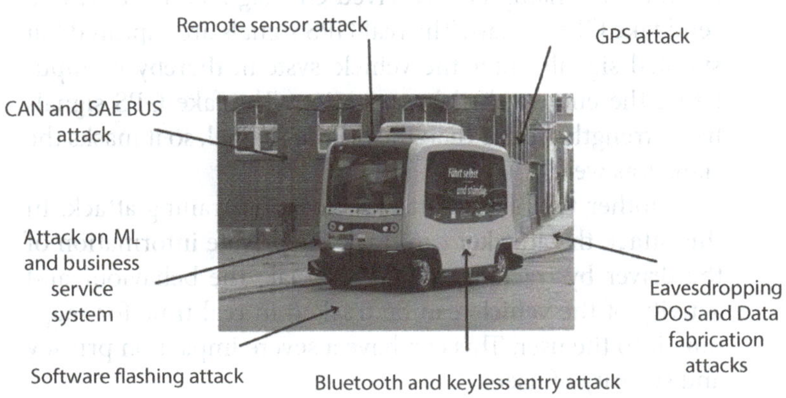

Figure 9.1 Demonstrating different cyber attacks.

9.4.1 Different Attack Surfaces and Resulting Attacks

- Bluetooth, TPMS and keyless access: These are some popular services used on autonomous vehicles. Bluetooth is used for smartphone connection and Bluetooth headset. Tire Pressure Monitoring System (TPMS) which tracks the pressure in tires and informs about low pressure to users. Keyless entry is a specialised feature of most modern cars and used by autonomous cars as well. It allows users to enter a car without inserting the key. They all use wireless technologies for propagating messages and operation. They are potentially vulnerable to proximity attacks.

  Where attackers can clone or even set up new wireless service and mimic existing networks properties to gain access. Direct wireless network penetration is also possible. Bluetooth is particularly targeted by hackers for cyber attacks.
- GPS: GPS stands for Global Positioning System. It is a satellite-based navigation system that provides geo-location and time-based location to GPS receivers anywhere on earth. Autonomous vehicles rely heavily on a GPS system as it is the basis for driverless navigation and path finding. Accurate road-map and location information is a must for AVs [12] to perform their task.

  A GPS system can also be compromised and there are attacks related to it as well. In a GPS spoofing attack, the attacker can manipulate received GPS signal arbitrarily and send fake GPS signals. The real GPS signals are captured but spoofed signals enter the vehicle system, thereby manipulating the entire vehicle's navigation. The fake GPS signals have strength greater than the original signal, so it masks the signals as well.

  Another GPS-based attack is location trailing attack. In this attack the attacker can track the private information of the driver by tracking the GPS signal. The behaviour and activity of the vehicle can be tracked in real time for mapping it to the user. This can have a severe impact on privacy and security of users.
- Lidar, Camera and Radar: These are the detection systems of an autonomous vehicle system. They help to find and map the surroundings for the car's internal systems for the

vehicle's mobility. Lidar sensors are great for autonomous vehicles as they measure distance to objects at high speed and with great precision. Unlike Radar that uses radio waves, Lidar uses light waves that bounce off the surface to calculate the distance.

These sensors are prone to remote sensor attacks where the adversary can remotely attack and take control of these sensors. Once controlled, they can have a devastating impact on the vehicle's navigation and totally disrupt it. They are part of core components that form the internal navigation system. Any disruption in that could have huge consequences.

- Machine Learning System: The machine learning system of an autonomous vehicle is what separates it from an automated vehicle. These systems are capable of learning in real time and changing navigation accordingly. Although they receive instructions from remote server regarding certain parameters, an ML system works locally alongside other detection systems. Any remote attack on core components can disrupt it and the car would lose all its navigational properties.
- ECUs: ECU stands for Electronic Control Unit. ECUs are embedded control units that are responsible for specific functions in a car. They receive inputs from different parts of the car and work in tandem for regulating specific actions. They can control gear shifts, ignition system, power transmission, temperature control, etc. Since they are embedded softwares and are reprogrammable, attacks like reverse engineering, remote code execution and fuzzing can introduce bugs and backdoors in them.

  The main problem with ECUs is Plashing attacks, where the remote system can be tricked into allowing attackers to flash the firmware and damage it. System vulnerabilities and unpatched softwares can provide access to the core system and damage it.
- Integrated Business Service: Integrated business services are quite essential for autonomous vehicle systems as many times the vehicle systems are directly connected to a manufacturer's back-end systems. The focus is providing customers with services like customer support, entertainment services, etc. Such services can lead to serious issues in

security. An attacker might gain access to client privileges and further escalate to higher ones to attack connected systems. Phishing attacks and code execution are a few ways to gain access to systems.

Embedded systems (ECUs) in an autonomous vehicle system pose a certain risk as if those essential units got hijacked or compromised then it would have a devastating impact on mobility and autonomy of the entire system.

- Infrastructure: Autonomous vehicles have communication and central infrastructure. The core components of the system form an essential part of that. If attackers gain illegal access to communication infrastructure, then they can cause multiple attacks. If ECUs and other components cannot communicate with each other, then the whole system will come to a standstill. Attackers would be able to exploit vulnerabilities in entities and cause the collapse of the total infrastructure.

9.5 Possible Attacks in Autonomous Vehicles

9.5.1 Internal Network Attacks

Existing internal network attacks include remote sensor attacks, GPS spoofing attacks, location trailing attacks, close proximity vulnerabilities, controller area network (CAN) and Society of Automotive Engineers (SAE) J1939 buses vulnerabilities, electronic control units (ECUs) software flashing attacks, and integrated business services attacks, which are explained below.

- Remote Sensor Attack

  In autonomous vehicle systems, different detection systems like Lidar, Radar, camera and other sensors communicate with each other and with the central system through an in-vehicle network. They utilize wireless technology systems for doing so. This can cause attackers or external entities to gain access or control these systems with enough tweaking or hacking. In comparison to a normal vehicle system where everything is hard-wired and connected with cable, autonomous systems use more wireless technology in that regard, making it vulnerable to sensor attacks.

Remote code execution can provide the highest level of system access if attackers find any vulnerability in the system. With access to sensors, ECUs can be controlled, which can pose a serious threat to navigation and control systems.

- GPS Spoofing attacks
  GPS (Global Positioning System) is a global satellite network that provides locations and time information of entity. As explained earlier, an attack surface regarding GPS can prove fatal. The attacker manipulates the received GPS signal arbitrarily [11]. The receivers are provided with fake GPS signals. The signal strength of fake signals overpowers the original ones, thus overshadowing it.

  The GPS information provided by an attacker can prove fatal as that would change the trajectory of the autonomous vehicles without the user's knowledge, leading to dangerous consequences.
- Close Proximity Vulnerabilities
  These vulnerabilities are caused by the accessing vulnerabilities in sensors. When attackers come in close proximity of the sensor, they can attack and take control of various sensors using different mediums like Bluetooth, in-vehicle networks, etc. Keyless entry system also falls under proximity sensors where the signals from the device locks or unlocks the vehicle.

  The Bluetooth systems are equally weak and prone to various kinds of attack like man-in-the-middle attacks, key spoofing, etc.

  Several cryptographic protocols have been devised for implementation on Bluetooth technology. But they fail in actual implementation due to their inefficiency. Further, these protocols are not made to prevent code injection attacks.
- Key Fob and Keyless Entry Attack
  A key fob is a small electronic device used to provide access to physical objects. They are also considered as hardware tokens providing on-device, one-factor authentication to facilitate access to a vehicular system.

  A key fob communicates through wireless modes and instructs the autonomous vehicles about action intended. An attacker can block or jam the fob signal, thereby preventing closing of doors.

Even more dangerous can be gaining control of other remote keyless entry keys. These keys can not only open and close doors but start the vehicle and allow access to other essential services. In some cases, the attacker may not be able to block the signal but can capture it and use it to replicate the original signal for gaining access. These attacks can be deduced as denial of service since the attack will not allow the legitimate user to accessing their own vehicle. Other store and modify attacks will not affect the users directly but can lead to theft or other monetary consequences.

- CAN and SAE J1939 Buses Vulnerabilities
  SAE1939 is defined as a standard for communication and transmitting diagnostic messages. It is a standard vehicular bus that operates between its components. Similarly, CAN (Controller Area Network) is a robust bus standard that allows micro-controllers and other components to communicate without centralised control. These bus standards also face security threats and attacks.

  The main problem with these buses is that they do not allow authentication and encryption [13]. Due to lack of proper security measures, attackers can connect illegal and unauthorised nodes to these networks to sniff out traffic. They might even be able to collect sensitive information regarding firmware and underlying software.

  The CAN bus network is also susceptible to DOS (Denial of Service) attacks. As the internal mechanism allows the node with highest priority to communicate first, an attacker with access to polling mechanism can increase their priority, thereby restricting others from communicating.

- Software Flashing Attacks
  Autonomous vehicles consist of small electronic units that work in tandem to make the whole system work together. These units are reprogrammable and are connected to the main bus for network level communication. A few of their operations are ignition control, temperature control and system monitoring. They even allow new firmware updates and enhance features.

  Attackers can leverage those features to critically attack the back-end firmware. With attacks on these electronic units, adversaries can flash the system [8] with corrupted software which can provide them control over the entire navigation system and even remote business services. Possible attacks can be code injection, fuzzing [14], plashing, etc. Each of these attacks have different attack surfaces and attack vectors.

Prominent among these is Plashing attack, which tricks the internal vehicular systems into installing software to its firmware. This can use unpatched vulnerabilities or other attacks as vectors to accomplish its task.

9.5.2 External Attacks

Autonomous vehicles not only communicate internally but deal with other external entities as well, like other vehicles, smart gadgets, connected smart components, etc. These entities open up space for new kinds of attacks where the attacker need not penetrate deep into vehicular system but can attack external facilities and components. Different wireless devices can lead to wireless attacks based on Wi-Fi, Bluetooth and radio communications. Such attacks are further decribed below.

- DOS attacks
  A DOS (denial of service) [15] attack can be stated as one of the major external attacks on autonomous vehicle systems. The denial to information is one of the major problems. The attacker can cut off legitimate signals from reaching the system, thereby making it useless. DOS can also be performed by jamming in useless noise in between legitimate signals and creating issues with network nodes.

  Autonomous vehicles depend upon various interdependent systems to function properly and in synchronization. Denial of these systems or incorrect reception from any system can have a cascading fallout effect. Delay in transmission can be a real challenge for real-time systems and can lead to accidents and major losses.
- Routing Attack
  Routing is the process of selecting an appropriate channel for traffic in a network or across networks. It ensures that the data packets are delivered efficiently and with minimal error. Routing attacks are tricky to deal with as adversaries use inherent problems and vulnerabilities in existing routing protocols to launch various attacks. The routed packets can be changed or made to fall in an infinite loop. Various attacks

related to these strategies are black hole attack, wormhole attack and grey hole attack. These attacks work differently in the way they are launched.

A black hole attack [16], [18] acts like a black hole and gobbles up each packet that is forwarded to it. It does not further transmit to another router, it simply deletes everything. A wormhole attack, as the name suggests, acts between two nodes. The attacker sets up a pathway for information tunnelling to its own selected end point. Grey holes are transient black hole attacks.

- Data Replay Attacks

  In a data replay attack [19], the attacker can store the packets or incoming information and later forward them to the source. It is maliciously repeated or delayed intentionally. The attacker intercepts the data by placing himself in between the communications between the nodes. It mainly occurs at the network layer of the system. The attacker can possibly read the entire communication and act as a mediator, which can hamper the data-flow.

- Data fabrication

  Data fabrication or spreading wrong data [20] is not a direct attack but ultimately has the consequence of an attack. Autonomous vehicles work on accurate and high-precision data from various sensors like Lidar, Radar and camera units. They help to map out the environment in real time for proper and safe navigation. If the data to the system is wrong or corrupt, that would have multiple levels of failure starting from navigation.

  Falsified data can not only disrupt the vehicle but also its total infrastructure and even back-end firmware. Connected vehicles may suffer from route guidance failure, traffic congestions and even accidents.

- Impersonation attack

  Autonomous vehicles have a unique identification number which is used for data transmission, information reception and utilizing various remote services. In this attack, the attacker impersonates or takes away identities of various devices and tries to spoof their messages. If GPS and other sensors' information is extracted, then it can cause mayhem. Impersonating the node can be done by using another or fake identity.

An attacker can impersonate different node points based on their vulnerabilities and location. If he spoofs just one node then it is a node impersonation attack. It is also possible to spoof large number of nodes simultaneously. Access to a large number of nodes can lead to DOS, MITM (man-in-the-middle), and data fabrication attacks.

- Eavesdropping attack
  An eavesdropping attack happens when an attacker tries to listen to communication from a medium that they should not be allowed to [6]. They can intercept, modify or delete data if provided with enough access and privileges to do so. Also known as sniffing and snooping, eavesdropping of autonomous vehicle communication infrastructure can lead to technical information which can later be used to craft an attack.
  These are a kind of passive attack that just monitors the system but does not actively attack it. The adversary can perform network traffic analysis and position tracking with this technique. This can harm the privacy of the users and integrity of the system.

9.6 Defence Strategies against Autonomous Vehicle Attacks

The abovementioned cyber attacks against autonomous vehicles can disrupt the entire industry if we do not plan for defences with the right measures. Wehether it is defence at source, intermediary node or internal network, it becomes critical considering the technology and human lives at stake. The following defence strategies describe measures for both internal and external network attacks.

9.6.1 Against Internal Network Attacks

Defence against remote sensor attack
Remote sensor attacks can be prevented by having proper authentication and consistency checks in place. Sensor protection units can be placed where attackers try to attack them. Physical layer protection to these devices can monitor and regulate attacks against them.

- Defence against GPS spoofing
  GPS spoofing can be prevented by adapting various methods and mechanisms that can cross-verify those details. Several strategies include velocity consistency checks, statistical checks, and GPS augmentation.
- Defence against close proximity attack
  To defend against close proximity attack, the channel or medium should have proper security measures in place. The devices should have encryption mechanisms and check sum function for packets to validate their integrity.
- Defence against keyless entry attack
  The user must be aware of people trying to snoop and collect keyless data. The attacker needs to be in close range for such an attack to work. Proper encryption of data and signals can prevent these attacks.
- Defence against CAN and SAE J1939 Buses vulnerability
  A few solutions can protect against CAN and SAE buses vulnerabilities like network segmentation, encryption and authentication. Network segmentation can segregate the electronic units into essential and non-essential, reducing their attack potential.
- Defence against software flashing attack
  In order to protect firmware from flashing attacks, hardware level access should be very tough to obtain. On top of that, encryption, authentication and integrity checks can help in securing further data from attackers. Reverse engineering should be made as complex as possible to be able to read and understand underlying mechanism.

9.6.2 Against External Attack

Protection against internal network attacks is not sufficient for the overall security of the autonomous vehicle system. The following ways illustrate defence strategies against external attacks:

- Against DOS
  Certain strategies like data rate limiting, bandwidth limiting, sliding mode and adaptive estimation can help detect and limit DOS attacks. Real-time detection and prevention is tough even for a good security system due to the nature of a DOS attack. Packet filtering and protocol management can also help.

- Against Routing Attacks
  Routing attacks can be defended by implementing various kinds of techniques like swarm algorithm of artificial intelligence, variable control chart, route optimization. Various researched and proposed algorithm are self-analysing and smart for route correction and prevention against common attacks and failures.
- Against Data Replay Attack
  Data replay attacks can be prevented by not allowing the intermediary to record and send the message. This can be done by timestamping each message, and out-of order timestamp would be discarded. Even single use can use random session key, and if the message does not reach its destination in the intended time or format, the session would expire, indicating foul play.
- Against Data Fabrication
  Data Fabrication can be prevented by ensuring the authenticity of the source's data. For false location data, location detection data can be used when in a connected network. A dynamic time window can help filter out forged data from falsified traffic, preventing major system disruption.
- Against Eavesdropping
  Eavesdropping is listening to unauthorised messages and signals. It can be prevented simply by implementing anonymity, trust-based systems and scheduling mechanisms. Using strong encryption and cryptographic protocols can help as well. Offloading data to other channels unknown to attackers can help as well.

9.7 Cyber Threat Analysis

In order to properly manage and mitigate cyber threats, various kinds of strategies and procedures can be implemented like threat categorization, vulnerability analysis, proper risk management and attack classification.

A well-established and properly regulated framework used for vulnerability assessment is CVSS (common vulnerability scoring system) [1, 3]. CVSS is one of the most reliable and advanced assessment programs that scans through a huge repository of threats and vulnerabilities and classifies them based on their characteristics and severity. It consists of three metric groups, namely Base, Temporal [20] and Environmental. They assign a score ranging from 0 to 10. It can stand as a well-suited standard measurement for industries, organizations and governments. Here, it finds application in Autonomous systems as well. Currently it follows up with a new version of v3.X improving on the previous standard.

9.8 Security and Safety Standards in AVs

Software defence strategies work well enough, but it will not help if things are not standardised or documented for everyone to follow. A few global organizations have introduced security standards and safety measures for autonomous vehicle systems. Those standards are as follows:

- ISO 26262:
  ISO 26262, titled "Road-vehicle-functional safety", is an international standard for operational safety of electrical and electronic systems that are installed in production road vehicles. It defines various components and their interaction with safety standard. It defines management of functional safety by categorizing various hazardous events into level of safety. It provides a Safety life cycle for a smooth and error-free development process. This life cycle ensures that specific safety requirements are met and global standards are being followed.

  It also constitutes ASIL (Automotive Safety Integrity Level) analysis. ASIL refers to an abstract classification of inherent safety risk in an automotive system or elements of such a system. ASIL classifications are used within ISO 26262 to express the level of risk reduction required to prevent a specific hazard, with ASIL D representing the highest hazard level and ASIL A the lowest.
- SAE J3061:
  Built on the previous standard of ISO 26262, SAE J3061 identifies security threats and attacks more vigorously. It properly lays guidelines for threats, vulnerabilities and attack methods on vehicular systems. It differentiates between system level security and system safety.

  It makes sure that proper security procedure is built and tested into each feature and component. It provides some tools and procedures for designing, developing and validating cyber-physical systems. On top of that it provides enough flexibility for organizations to build further standards on top of it. General guidelines and instructions are also provided for autonomous systems.

9.9 Conclusion

This book chapter about cyber security and threat analysis in autonomous vehicles provides all-round information regarding the threat landscape, cyber attacks and defence strategies for mitigating the same. Various attacks were well discussed related to internal and external network threats. The international standards defined can help reduce the occurrence of security flaws and bugs at source if properly implemented.

The attack methods and surfaces defined are very operational and are being exploited everywhere. People should be made aware of the cyber security scenario and its current capabilities in dealing with such problems. The weakest link in the security scenario is the human one. Various attacks take the human factor into consideration.

Overall, these attacks and their vectors are the result of various flaws that can creep into the system during its development, designing or testing phases. Some flaws are due to the inherent nature of that system. So, proper all-round system testing, and vulnerability assessment is a must for autonomous vehicles. With manual intervention in mobility being impossible, manufacturers need to be especially wary of cyber threats and attacks targeting them.

References

1. Y. Fu, F. R. Yu, C. Li, T. H. Luan, and Y. Zhang, "Vehicular blockchainbased collective learning for connected and autonomous vehicles," *IEEE Wireless Commun.*, vol. 27, no. 2, pp. 197–203, Apr. 2020.
2. Zack Whittaker,"Security bugs let these car hackers remotely control a Mercedes-Benz", *techcrunch*, 2020.
3. A. Nanda, D. Puthal, J. J. P. C. Rodrigues, and S. A. Kozlov, "Internet of autonomous vehicles communications security: Overview, issues, and directions," *IEEE Wireless Commun.*, vol. 26, no. 4, pp. 60–65, Aug. 2019.
4. Yağdereli E, Gemci C, Aktaş AZ. A study on cyber-security of autonomous and unmanned vehicles. *Journal of Defense Modeling and Simulation*. 2015;12(4):369-381.
5. A. Palanca, E. Evenchick, F. Maggi, and S. Zanero, "A stealth,selective, link-layer denial-of-service attack against automotive networks," in *Proc. Int. Conf. Detection Intrusions Malware, Vulnerability Assessment*, 2017, pp. 185–206.

6. S. Balakrishnan, P. Wang, A. Bhuyan, and Z. Sun, "Modeling and analysis of eavesdropping attack in 802.11ad mmWave wireless networks," *IEEE Access*, vol. 7, pp. 70355–70370, 2019.
7. B. Sheehan, F. Murphy, M. Mullins, and C. Ryan, "Connected and autonomous vehicles: A cyber-risk classification framework," *Transp. Res. A, Policy Pract.*, vol. 124, pp. 523–536, Jun. 2019.
8. R. R. Brooks, S. Sander, J. Deng, and J. Taiber, "Automobile security concerns," *IEEE Veh. Technol. Mag.*, vol. 4, no. 2, pp. 52–64, Jun. 2009.
9. W. Xu, A. Willecke, M. Wegner, L. Wolf and R. Kapitza, "Autonomous Maneuver Coordination via Vehicular Communication," *2019 49th Annual IEEE/IFIP International Conference on Dependable Systems and Networks Workshops (DSN-W)*, 2019, pp. 70-77, doi: 10.1109/DSN-W.2019.00022.
10. Barabás *et al.* 2017 IOP Conf. Ser.: *Mater. Sci. Eng.* 252 012096
11. J. Nunez, V. Tran, and A. Katangur, "Protecting the unmanned aerial vehicle from cyberattacks," in *Proc. Int. Conf. Secur. Manage.*, 2019, pp. 154–157.
12. C. W. Axelrod, "Autonomous vehicles meet inhospitable roadways," in *Proc. IEEE Long Island Syst., Appl. Technol. Conf. (LISAT)*, May 2019, pp. 1–6.
13. M. Bozdal, M. Samie, and I. Jennions, "A survey on CAN bus protocol: Attacks, challenges, and potential solutions," in *Proc. Int. Conf. Comput., Electron. Commun. Eng. (iCCECE)*, Aug. 2018, pp. 201–205.
14. D. R. Jeong, K. Kim, B. Shivakumar, B. Lee, and I. Shin, "Razzer: Finding kernel race bugs through fuzzing," in *Proc. IEEE Symp. Secur. Privacy (SP)*, May 2019, pp. 754–768.
15. S. Kumar and K. S. Mann, "Prevention of DoS attacks by detection of multiple malicious nodes in VANETs," in *Proc. Int. Conf. Automat., Comput. Technol. Manage. (ICACTM)*, Apr. 2019, pp. 89–94.
16. S. S. Albouq and E. M. Fredericks, "Lightweight detection and isolation of black hole attacks in connected vehicles," in *Proc. IEEE 37th Int. Conf. Distrib. Comput. Syst. Workshops (ICDCSW)*, Jun. 2017, pp. 97–104.
17. J. Wang, C. Jiang, K. Zhang, T. Q. S. Quek, Y. Ren and L. Hanzo, "Vehicular Sensing Networks in a Smart City: Principles, Technologies and Applications," in *IEEE Wireless Communications*, vol. 25, no. 1, pp. 122–132, February 2018, doi: 10.1109/MWC.2017.1600275.
18. K. C. Purohit, S. C. Dimri, and S. Jasola, "Mitigation and performance analysis of routing protocols under black-hole attack in vehicular ad-hoc network (VANET)," *Wireless Pers. Commun.*, vol. 97, no. 4, pp. 5099–5114, Dec. 2017.
19. A. Mondal and M. Jana, "Detection of fabrication, replay and suppression attack in VANET—A database approach," in *Proc. Conf. Advancement Comput., Commun. Electron. Paradigm*, 2019, pp. 38–42.
20. R. M. Shukla and S. Sengupta, "Analysis and detection of outliers due to data falsification attacks in vehicular traffic prediction application," in *Proc. 9th IEEE Annu. Ubiquitous Comput., Electron. Mobile Commun. Conf. (UEMCON)*, Nov. 2018, pp. 688–694.

10

Big Data Technologies in UAV's Traffic Management System: Importance, Benefits, Challenges and Applications

Piyush Agarwal, Sachin Sharma and Priya Matta*

Department of Computer Science and Engineering, Graphic Era (Deemed to be University), Dehradun, India

Abstract

The Internet of Things (IoT) has gained widespread adoption, prompting the addition of new connected devices to the network, such as Unmanned Aerial Vehicles (UAVs). UAVs provide a host of social benefits, including improved personal safety, security, and productivity. These help first responders save lives during natural catastrophes, find missing children, and combat wildfires. They assist the Coast Guard in rescue missions and the Border Patrol in maintaining national security. UAVs provide a practical solution to the limitations of existing terrestrial IoT infrastructure, as well as new ways to deliver IoT services through a wide range of applications. UAVs are well-suited to collecting aerial data for constant use in large industrial facilities, allowing for speedy and seamless data collection to support informed business decisions. By eliminating the requirement for a drone operator, an automated UAV system improves efficiency while also providing continual, frequent, and real-time data access. A UAV system collects data from the air and is a useful tool for organizations automating large-scale activities. With these benefits, it is expected that the number of UAVs flying in low-altitude airspace above congested cities will rise. As a result, a system for monitoring and managing UAV traffic must be put in place. This adds to the difficulty of doing research, such as guaranteeing the safe operation of UAVs in high-traffic regions. This research looks into UAV Traffic Management (UTM), with a focus on big data approaches in UTM. The benefits, limitations, and implementation of big data approaches in UTM systems are also discussed in this chapter. This chapter also looks at the

**Corresponding author*: mattapriya21@gmail.com

Romil Rawat, A. Mary Sowjanya, Syed Imran Patel, Varshali Jaiswal, Imran Khan and Allam Balaram (eds.) Autonomous Vehicles Volume 1: Using Machine Intelligence, (181–202) © 2023 Scrivener Publishing LLC

results of recent study in the same domain. Finally, the chapter concludes with a consideration of the traffic control system for UAVs in the future.

Keywords: Big data, drones, Internet of Things, traffic management system, UAV, UTM

10.1 Introduction

In recent years, there has been a greater emphasis on the deployment of UAVs in low-altitude airspace. Several operators use UAVs to reach a specific location because of the limited capacity of UAVs. As the usage of UAVs in the airspace grows, so does the chance of collision, which is never acceptable. The services provided by such a system would benefit stakeholders such as the International Civil Space Organization, the National Civil Space Airspace Authority, aircraft manufacturers, air navigation service providers, air traffic controllers, pilots, unmanned aerial system service suppliers, and the general public. As a result, a traffic management system that controls UAV traffic is required. NASA and the Federal Aviation Administration (FAA), the US civil aviation regulatory agency, began collaborating on "The Next Era of Aviation" in 2015. "When it is proposed to have the first autonomous car in the sky," was the topic of discussion for the proposed plan. According to Rumba *et al.* [1] there was only one company running fully automated UAVs in Africa until mid-2019. Since 2017, aircraft manufacturers have been using the idea of Urban Air Mobility (UAM) for operational demonstrations and plans for the first flight. The current approach for a future generation autonomous air transportation system is unsuitable; the current solution uses voice commands to handle air traffic. In addition, the current air space management system has set aside the majority of the available airspace. To cope with the tenfold rise in flight operations by 2035, where autonomous flights will account for 95 percent of all flights, a system that can govern the air space as well as manage autonomous flight autonomously is required.

Various sensors gather a vast quantity of data during the flight, including humidity, temperature, video, thermal image, and so on. The current high-definition video encoder has a coding rate of 30Mbps. The data collected by sensors on a single UAV can be as high as 5Mbps. The data generated by a UAV swarm with a size of around 30 will be around 150Mbps. As the number of sensors, their resolution, and accuracy increase, so does the magnitude of the data; such massive data is referred to as big data.

A literature review of recent work on path planning and object detection avoidance for UTM is covered in this chapter. This chapter then goes through the basics of UTM, including tracking and tracing, path suggestion, path deviation, path suggestion, collision detection, collision avoidance, and so on. The importance of big data technologies is discussed in the next section, followed by its benefits and challenges. This chapter will compare and contrast the work done in the same area later on. The chapter concludes with a case study on Prime Air, an aerial delivery service, as well as a discussion on the future potential of a big data-based UTM system.

10.2 Literature Review

This section provides a brief overview of work in the fields of UAV path planning, object recognition, and avoidance in recent years. Yadav *et al.* [2] proposed the architecture for the UAV traffic management system (UTM) which can register and provide the credentials to the UAV and pilots. Before the flight, the path or trajectory has to be registered to the system. The path trajectory of all the flights is combined and checked for the possibility of collision. The alert will be issued in the case of deviation from the trajectory and also the path of the unauthorized UAV is predicted. Ho *et al.* [3] proposed the extended formulation of multi-agent pathfinding that was the extension of the conflict-based search (CBS) and enhanced CBS (ECBS). The proposed work introduces heterogeneous agents with computation geometry and Spatio-temporal pruning is used to reduce the search space. Also, they have included batching method along with cooperative A* into the CBS and ECBS that enhances the number of requests addressed by it efficiently.

Lin *et al.* [4] proposed a UAV flight surveillance system based on the hierarchical UAV traffic management system. The proposed system adopts the general ADB-S communication infrastructure. The system will accept UAV flight requests after registration verification of the UAV and pilot, then the system will proceed to the flight plan. The system also consists of the cloud used for flight surveillance data accessing and exchanging. The system uses Detect and Avoid function to calculate the time to collide when two UAVs are approaching. The collision alert will be sent to the mobile of the pilot through voice call. Airspace was shared by the different UAV service providers, and to control all the UAVs using the centralized UAV traffic management is a difficult task. Ho *et al.* [5] proposed a decentralized system in which each service provider will resolve the conflict within its system as well as with other service providers. The proposed system uses the Preflight CDR (conflict detection and resolution) method to provide conflict-free paths to UAVs before they take off. The pre-Flight CDR problem was resolved by sending a multi-agent path-finding framework. In the case of any conflict, the proposed system uses the given two algorithms: i) resolution with prioritization as a baseline method and 2) resolution with negotiation. Using these algorithms each service provider will identify the effective cost and the path of UAVs with lesser cost will be changed.

For UAV-based monitoring of nuclear power plants, Kliushnikov *et al.* [6] propose the set of route-based reliability models. The proposed model presents the approaches to calculating the probability of the successful fulfillment of the plan to cover the monitoring station and nuclear power plant by UAV. The model keeps account of the failure of UAV and utilization features of redundant UAVs.

Wan *et al.* [7] proposed a mobile edge computing-based online data processing network. The proposed network consists of three layers. The bottom layer consists of distributed sensors that generate the raw data. UAV-Base stations are placed at the middle layer that collects the raw data from the bottom layer and performs the initial processing. Finally at the top layer cloud receives the processed data for final computation. Lyapunov optimization-based scheduling algorithm is used for edge processing. For hovering UAV-Base stations deep reinforcement learning is used as the path planning algorithm and CNN is used to produce the reward of the candidate action.

For object detection and avoidance Smolyanskiy *et al.* [8] proposed a system TrailNET for UAVs that is based on a deep neural network (DNN) technique. For the improvement of accuracy and efficiency rectified linear units activation functions are used. For object detection and avoidance, vision processing is performed using a DNN node that uses TensorRT, YOLO DNN is used for object detection, and DSO visual odometer algorithm for converting the output to a camera-centric depth map. Desired movement command is computed by the controller node as per DNN predictions, object detection and avoidance, and teleoperation commands. Energy is one of the most crucial aspects of UAVs. For efficient utilization of the battery in a UAV while on a mission, it should use an optimized path with smooth curves. Fu *et al.* [9] proposed a similar kind of work; in the method, they proposed three functions to optimize the battery life. The proposed cost functions are path security cost, length cost, and path smoothness cost. Path feasibility is determined by the path security cost, whereas the cost of energy consumption can be optimized by the length cost and smoothness cost functions. To set the operation such as substitution operation, cross operation, mutation operation length operation, and smoothness operation, a heuristic evolutionary algorithm is proposed that generates the effective path.

Monwar *et al.* [10] worked with a similar aim and proposed the framework. The proposed framework includes the energy required for hovering and data transmission while flying. The proposed algorithm also takes care that the trajectory or path followed by the UAV does not contain any cycle. For real-time path planning, Zhan *et al.* [11] proposed A* based algorithm for UAVs in a 3D large-scale battlefield environment to increase the survival rate and low fuel consumption. Given the two points, A* based algorithm provides the optimized path considering the factors such as altitude, steepness of the path, a radius of the turns.

10.3 Overview of UAV's Traffic Management System

With the increased use of unmanned aerial vehicles (UAVs) in the airspace, there will be more air traffic, which will raise the possibilities of a collision and make it more difficult to control aircraft, whether autonomous or human. The traffic management system for airspace can be utilized to overcome or lessen these issues. Figure 10.1 depicts some of the characteristics of a UTM system.

- Tracking and tracing: For surveillance and security purposes, it is critical to track and trace every UAV flight. This function aids in determining if the UAV is on the right track or not.
- Path Suggestion: The path is one of the most important aspects of any flight. The flight path must be updated for a variety of reasons, including before the flight takes off, a conflict in the flight path with another flight, and so on. Weather conditions or other factors can cause a flight to divert from

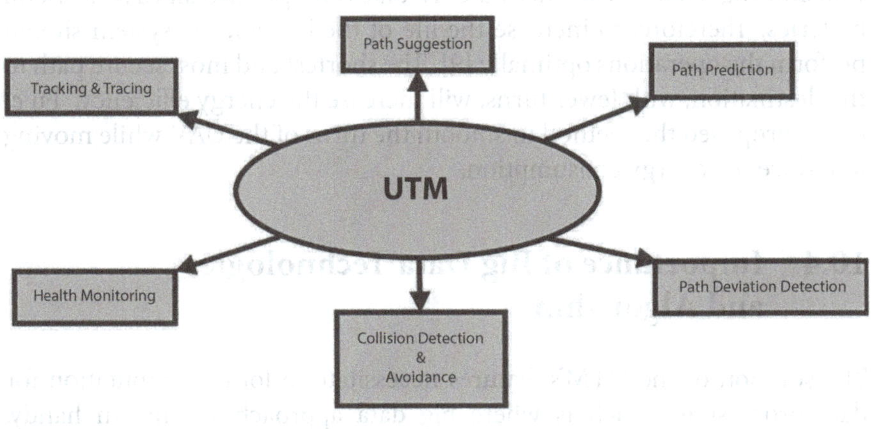

Figure 10.1 Characteristics of UTM.

its intended path. Also, in real time, any bird or other unauthorized UAV may fly across the path of the UAV; to avoid a collision, the UAV must deviate from its path; in this situation, the system should offer or recommend a different route for the UAV to reach its goal.

- Path Prediction: In the open air any unauthorized UAV can take up a flight, and its path can be predicted based on the trajectory it is following. This path prediction can reduce the chances of collision with the authorized UAV.
- Path Deviation Detection: The UTM should have the feature to detect the path deviation of the flight to the safety and security of the deviated UAV as well as for the other UAVs.
- Collision Detection and Avoidance: The collision of the UAV with another UAV or any other object may result in the UAV crashing and also may permanently damage the UAV. Therefore, to overcome this condition the system should include the functionality of predicting the chances of collision and provide the suggestion to avoid it.
- Health Monitoring: Health of an UAV includes energy remaining, working condition of its propellers, their motors, sensors working conditions, and network connectivity. For better efficiency of the system, it should provide these features. Energy is the most crucial part of any UAV; therefore, the system must continuously monitor the energy of the UAV so that it does not crash in the middle of its flight.

Every operation performed, such as moving, turning, hovering, sensing, transmitting data, and so on, by a UAV takes energy; also all UAV work on batteries. Therefore, to increase the life of the battery, the system should perform the operations optimally [9]. The shortest and most secure path to the destination, with fewer turns, will increase the energy efficiency. Fu *et al.* [9] proposed the method to smooth the turns of the UAV while moving to reduce the energy consumption.

10.4 Importance of Big Data Technologies and Algorithm

The solution of the UTM's features necessitates a lot of computation for data processing, which is where big data approaches come in handy. Artificial intelligence, blockchain, augmented reality, and cloud computing are some of the additional technologies that can be applied in traffic management systems [18]. Figure 10.2 shows some of the strategies that can be employed in UTM to improve the efficiency of the UAV and the system.

- Deep neural network (DNN): DNN has many layers of data processing between input and output layers. DNN can be used for object detection and identification from the image captured by the camera installed in the UAV.
- Fuzzy Logic: To handle uncertainty, a soft computing method called fuzzy logic can be applied. The value of uncertainty or membership degree lies between 0 and 1. Where 0 and 1 mean that is a member is not the part of the fuzzy set and a member is the set of the fuzzy set, respectively. A fuzzy set is a set on which fuzzy rules are applied to get the desired solution. Adhikari *et al.* [17] use fuzzy logic to find the optimized path. An optimized path will help the on-time completion of the task with the least risk and effective cost. Optimized also ensures the least energy consumption.
- R-CNN: It is a machine learning model used for computer vision especially for object detection. In UTM objecting detection and avoidance is one of the major issues, as a collision of the UAV with any other object can damage the UAV and the damage can be permanent or temporary [20].
- Single-shot multibox detector (SSD): The performance of SSD is better than that of R-CNN. This is because from SSD bounding box is eliminated and progressively decreasing convolutional filter is included.

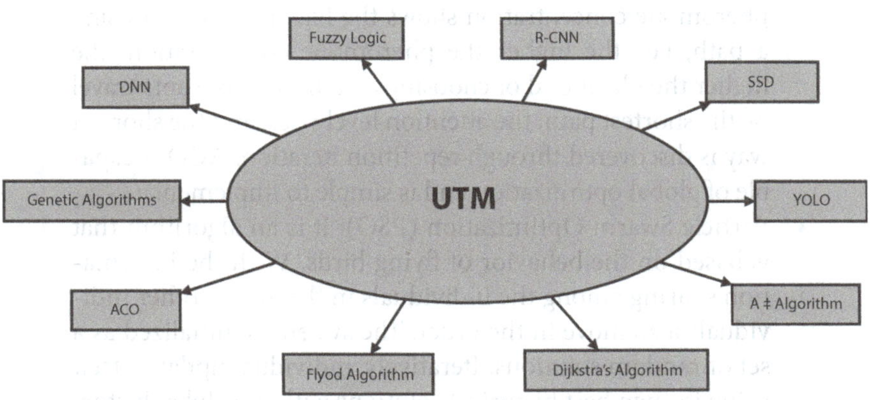

Figure 10.2 Big data techniques in UTM.

- YOLO: It is a neural network-based algorithm for object detection [19]. It detects the object in single forward propagation neural network and has excellent learning capabilities.
- A* Algorithm: With heuristic functions, the A* algorithm calculates the value of each neighboring node and compares their values. The node with the lowest value will be selected and this process continues till the selected node is the target node. The main advantage of this algorithm is that it has good real-time performance with few extension nodes. The drawback of this algorithm is that the size of the moving object is ignored.
- Dijkstra's Algorithm: Dijkstra algorithm provides the shortest path to the destination from source to destination. To obtain the shortest path the algorithm traverses from the source location to the destination by comparing node values and includes the node into the tree with the lower node value. The algorithm has a high success rate but it has a major disadvantage in that it traverses all the nodes to get the shortest path.
- Floyd Algorithm: Floyd algorithm works on the weighted graph to get the shortest path between the start and end locations. The weighted matrix is generated based on the edge weights of the graph and then finds the shortest path with distance. For big data calculation, this algorithm is not suitable due to high time complexity.
- Ant Colony Optimization (ACO): The ACO algorithm works in the same way that ants do when they're looking for food. When one ant is looking for food, it leaves a pheromone concentration so that other ants may smell it and follow it. The pheromone concentration shows the likelihood of choosing a path, i.e., the higher the pheromone concentration, the higher the likelihood of choosing a path. As more ants travel by the shortest path, the attention level rises, and the shortest way is discovered through repetition iteration. ACO is capable of global optimization and is simple to implement.
- Particle Swarm Optimization (PSO): It is an algorithm that is based on the behavior of flying birds. With the information sharing among the individuals in the group, other individuals also move in the order. The system is initialized as a set of random solutions. Iteratively, individual updates itself using its own best historical solutions with the global historical optimal solution to get the optimal value.

- Genetic Algorithm: Genetic algorithm provides the simulation solution for the natural selection of Darwin's biological evolution theory. It also provides the genetic evolution process of genetic mechanisms. For the start, the population having potential solution are coded, and the adaptive value of the individual is calculated. Selection, cross over and mutation is performed on the individuals using the principle of survival of the fittest, and the unfit individuals are eliminated. This process is repeated until the best individual is produced. This algorithm is combined with other algorithms and provides its iterative advantages.
- Neural Network Algorithm: Neural Network consists of the adjustable neurons and connected weights. The neural network has the capability of large-scale parallel processing, storage of distributed information, self-organizing, and self-learning. In path planning the calculation of the environmental condition and the collision, energy is done by the neural network. The optimized objective function is the sum of the distance function and the collision energy. The optimal path can be set by the iterative path point using the equation of motion [23].

10.5 Benefits of Big Data Techniques in UTM

The UTM system will be more efficient and accurate if big data techniques are used with it, as the majority of the procedures will be automated. Figure 10.3 depicts some of the advantages of the technology discussed in the preceding sections.

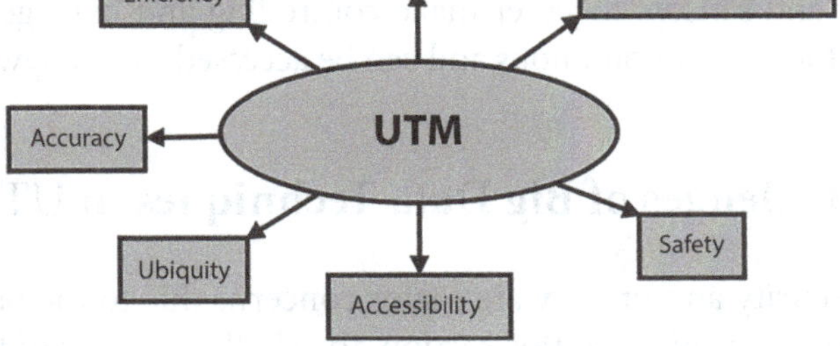

Figure 10.3 Benefits of big data technologies in UTM.

- Accuracy: Since all the UAVs are connected to the system, they can share real-time information with accurate location. Also, since the UAVs are controlled autonomously, the chances of error are negligible [24].
- Efficiency: The manual processing of path planning and object detection for a UAV system is not an easy task and is non-efficient. With the increase of the load, the chance of error also increases. Using the big data techniques these processes can be automated. As there will be no involvement of the human in any of these processes the chances of error will decrease and hence the efficiency of the system will increase [25].
- Reliability: The path provided by big data technologies makes sure that the UAV reaches the destination safely; also depending on the system, UAV can change its path to prevent the collision with the object or any other UAV [26].
- Quality of service: Path planning and collision detection are time-consuming and inaccurate if done manually. The likelihood of a collision can be anticipated in time using the approaches mentioned, and the real-time course can be altered to prevent the collision. Big data approaches leverage geographical data to accurately position the UAV as well as the destination and control every activity autonomously for path planning [22].
- Safety: The operations like turning, hovering, moving in a path, and deviating from the path are performed according to the pre-defined planning. This planning provides safety to the device as well as the item it is carrying [21].
- Accessibility: There are places where UAVs cannot be manually controlled; in those places these techniques will provide accessibility. This feature helps in disasters like floods, forest fires, and so on [27].
- Ubiquity: Big data techniques along with network connectivity and sensors together make controlling and management of the UAV autonomous and can be accessed from anywhere.

10.6 Challenges of Big Data Techniques in UTM

In UTM, security and privacy are major concerns due to the nature of its wireless communication. In this section, the challenges related to security, privacy, feasibility are discussed.

- Privacy: During the flight of a UAV with an embedded camera, there will be privacy issues for the people around it; also, there is some restricted or private area that does not allow any unauthorized object to enter the area. Therefore, the UAV must not fly over it. Also, the communication between the infrastructure and UAV is secured and protected.
- Threats and vulnerabilities: UAVs and drones can be viewed as viable and vital threats to information security. Due to design issues, some of the UAVs are having weak wireless security protection and footage encryption.
  - Prone to Spoofing – Serial ports are used for the transmission of the data to and from the UAV and due to its weak communication nature, the data in most cases is not encrypted. According to [12, 13] GPS Spoofing can be used to capture, modify, and inject information.
  - Prone to malware infection – UAVs are enabled with insecure communication protocols for controlling them through smartphones, tablets, laptops, or PCs [14, 15]. By creating and injecting the reverse-shell TCP payload into the UAV, the hackers can install malware on systems.
  - Prone to Data Interference and interception – Open non-secure wireless transmission is used for telemetry feeds for monitoring the UAV, as well as for the transfer of information. This non-secure transmission is vulnerable to threats such as data interception, malicious data injection, and alteration of the flight path.
  - Prone to manipulation – UAVs fly on a pre-programmed and pre-defined path, using radiofrequency and GPS spoofing; an attacker may take control of the UAV using a dummy signal or jamming it to crash it.
  - Prone to technical issues – UAVs may face technical issues related to application such as connection failure between UAV and the system, or application error that results in UAV crash or fly away. One of the major issues other than the application is the battery life; limited battery reduces the flying time. Also, in the cold weather the battery life is reduced, resulting in less flying time.
  - Prone to operation issues – If not controlled properly, the UAV/drone may cause damage or injuries to properties or personnel. These devices are very sensitive; a small accident such as being hit by a small object may result in

a crash or damage the rotaries of the UAV. Due to this it is very difficult to control the device and this can lead to a crash.

- o Prone to natural issues – Due to the weight of the drones, they cannot withstand wind. Extreme heat or temperature may result in a crash due to engine failure. Battery explosion also causes serious damage to the UAV. Since UAVs are not waterproof, they are incapable of flying in rain and if they crash into a lake, river, or any other water body they may immediately stop working.

- Surveillance: For air traffic control and flight information service it is important to have the information of all the flying objects. These flying objects cannot be detected by the primary and secondary radars as these radars are used to detect conventional aircraft with large sizes and mass. UTM consists of UAVs which have a small size and less weight. Also, material used for constructing the device cannot reflect the radar signals from the secondary radar of air traffic control This restriction does not apply to the larger objects with weight over 20-30 kgs, as these objects have bigger and larger reflective surfaces. Also, UAVs use low-level air space from 10 meters to a maximum of up to 200 meters [16].

10.7 Applications of Big Data Techniques in UTM

Many UTM applications can make use of big data approaches for citizen security and military needs. This section goes over a few of the applications (Figure 10.4).

- Natural Disaster Response and Control: In natural calamities such as earthquakes, tsunamis, and hurricanes, the only means to carry out rescue operations is by air. UTM can assist during this period by providing real-time information on the situation, the amount of people who need to be rescued, medical needs, and so on. Using UTM, UAVs can reach the afflicted area and offer photos of the area, giving rescuers precise information about the many requirements, such as the number of people who need to be rescued, medical needs, and food supplies. Additionally, photos of nearby

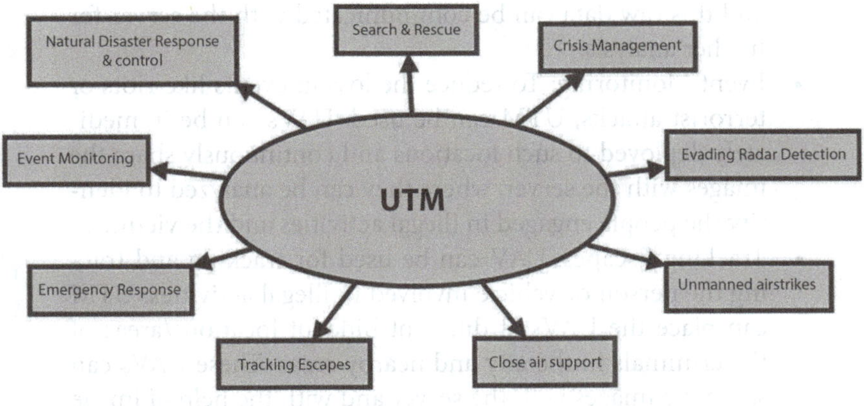

Figure 10.4 Applications of big data techniques in UTM.

unaffected places can be given to indicate alternative routes to the location.

- Search and Rescue: One UAV is not enough for search and rescue operations in a large area like a forest or sea, where it is dangerous. In this case, UTM provides the solution with the least number of UAVs and maximum area coverage. Each UAV can communicate to the server and share the images which in turn apply the object detection algorithm to find the people.
- Crisis management: UAVs can act as the base station to provide the connectivity with the limited functionalities, which allows a text message to be shared with the controller to get the ground reality. UTM can provide the path to the desired location to provide coverage to the maximum number of victims.
- Emergency response: UAV can help to deliver the medical support and food supply to the person who is out of reach or isolated. UTM can provide the path to the destination/patient and the patient can provide the health report that can be sent to the server for further analysis using big data techniques.
- Environmental Management: UTM can be used for measuring the air quality of different parts of the city. UTM can provide a safe path to cover maximum distance with a limited power supply. Using sensors, air quality can be measured,

and this raw data can be communicated with the server for further analysis.

- Event Monitoring: To reduce the loss in events like riots or terrorist attacks, UTM can be used. UAVs can be immediately deployed to such locations and continuously share the images with the server, where they can be analyzed to identify the people engaged in illegal activities and the victims.
- Tracking Escapes: UAV can be used for tracking and tracing the person or vehicle involved in illegal activities. UTM can place the UAVs at different hideout locations/areas of the criminals in the city and nearby areas. These UAVs can share the images with the server and with the help of image processing and computer vision technology the person or vehicle can be located.
- Close air support (CAS): UTM can be used in defense for CAS. UAVs that are equipped with weapons can be sent to the desired location using a path planning algorithm, and the UAV can send the images of the location to identify the locations of the enemy. The server can command the UAV to shoot at the enemy.
- Aerial surveillance/Reconnaissance: UAVs can also be used for gathering information related to the identification and tracking of military camps, weapons, vehicles, vehicle movement, improvised weapons, and so on. Using big data techniques in UAV traffic, the information gathered by the UAV can be analyzed and also provide the exact location with their number.
- Unmanned airstrikes: UAVs have been used to perform unmanned airstrikes against targets in the West Bank and military installations in Iraq and Syria. UTM airstrikes can be made automated, which will make the attack more successful.
- Evading Radar-Detection: As explained previously, UAVs are not detectable by radar systems; therefore UTM can be used to evade hostile territory and assess their strengths and weaknesses without coming into contact with their radar system.

In recent years, some work has been done on UTM systems based on big data technology. Table 10.1 lists the features and restrictions.

Table 10.1 Analysis of UAV traffic management system.

References	Algorithm	Architecture	Features					Limitations/future scope
			Collision avoidance	Path suggestion	Object detection	Energy efficient	Restricted area management	
[2]	Collision Detection	N/A	✓	✓	✗	✗	✓	This work can be extended for the unrestricted area where the registration can be done dynamically using the neighboring airspace data. In the case of deviation from the path, the new path to the destination could be suggested. Collision detection can be done using big data technique.
[3]	Binary Constraint Tree and A*		✓	✓	✗	✗	✗	A solution for detection and solution for the UAV that has deviated from the path.
[4]	Collision avoidance	Hierarchichal	✓	✗	✗	✗	✗	In the case of deviation from the path, the new path to the destination could be suggested.
[5]	Multi-agent path finding	Decentralized Architecture	✓	✓	✗	✗	✗	Due to path deviation during the flight, there are chances of conflict to arise. This condition can also be addressed here.

(Continued)

Table 10.1 Analysis of UAV traffic management system. (*Continued*)

References	Algorithm	Architecture	Features					Limitations/future scope
			Collision avoidance	Path suggestion	Object detection	Energy efficient	Restricted area management	
[6]	Routing	N/A	×	✓	×	×	×	The same work can be extended for monitoring station elements such as UAV Fleet, communications, crisis center.
[7]	CNN, Deep reinforcement learning, Lyapunov optimization	Layered architecture	×	✓	×	×	×	Path deviation is possible due to environmental conditions which may lead to a collision. This work can be further extended to resolve this issue.
[8]	DNN, YOLO DNN, DSO visual odometry algorithm	N/A	✓	✓	✓	×	×	The work of object detection and avoidance can be used for UAVs flying at a higher altitude than the proposed.
[9]	heuristic evolutionary	N/A	✓	✓	×	✓	×	This work can be extended for real-time object detection and avoidance.

(*Continued*)

Table 10.1 Analysis of UAV traffic management system. (*Continued*)

References	Algorithm	Architecture	Features					Limitations/future scope
			Collision avoidance	Path suggestion	Object detection	Energy efficient	Restricted area management	
[10]	Optimized path planning	N/A	×	✓	×	✓	×	Object detection and collision scheme is not considered.
[11]	A*	N/A	×	✓	×	✓	×	Object detection and collision scheme is not considered.

10.8 Case Study and Future Aspects

Amazon's Prime Air is a drone-based delivery service. Amazon gained approval for the prototype's testing in March 2015, under a waiver to the regulations in place at the time. The company will receive its FAA Part 135 air carrier registration in August 2020. The drone weighs about 55 lb (25 kg), and the company opted to give air delivery service within a 10-mile (16-kilometer) radius of its warehouse or facility centre at a speed of 50 mph (80.5 km/h), carrying up to a 5-pound (2.26 kilograms) packet. The UAV that the company uses for the service is efficient, steady, and safe. It can take off and land vertically, with a smooth transition between vertical and airplane mode. While in transition, sensors and algorithms such as multi-view stereo vision are used to detect stationary objects, while computer vision and machine learning algorithms are utilized to detect moving items such as paragliders or helicopters. A modest area with no people or obstacles is required for the UAV to land while delivering the product. People, animals, and other barriers are detected using explainable stereo vision and powerful AI algorithms. Detecting electric lines, telephone wires, or clotheslines in the customer's yard is the main issue. The computer vision approach was created by Amazon so that the UAV can avoid them as it descends into or ascends out of the yard. Amazon aims to achieve shipment zero which means "shipment with net-zero carbon". By 2030, the corporation wants to achieve net-zero emissions in 50 percent of its exports.

The internet of things, robotics, cloud computing, big data, and autonomous systems such as road traffic management systems, or UTM, have all advanced dramatically in recent years. These technologies provide new possibilities for developing autonomous applications that make better use of resources and are more efficient. UTM is compatible with smart city applications and can be employed in a variety of defense applications. In a smart city, a UTM system combined with big data approaches can be utilized to ensure inhabitants' safety and security. The application can be used to track and trace a vehicle or people who are involved in unlawful actions, or it can be used to monitor various occurrences [28]. The security personnel's efforts will be reduced, and the system's efficiency will improve.

The same technology can be utilized for military purposes to reduce losses at international borders and to get precise information about adversaries. Using big data technology in UTM, military activities such as close air support and aerial surveillance can be made more effective. The position of the adversaries can be determined using big data technologies.

It is possible to create a system that can distinguish between enemies, citizens, and security personnel from one's own country.

10.9 Conclusion

With the rise in popularity of UAVs or drones in the world of aviation, the autonomous aerial vehicle has become more popular in domains ranging from civilian to military. Big data-based UTM delivers economic, industrial, and commercial benefits because of its self-contained, simple to use, and adaptable nature. Tracking and tracing, path suggestion, collision detection and avoidance, path prediction of unauthorized UAVs, and other characteristics of a UTM system were described in this chapter. We have discussed some of the numerous big data strategies that can be used in the UTM system. After discussing the benefits of big data technology in UTM, this chapter discussed how data transmission from UAV to server and among different parts of the system leads to security, safety, and privacy concerns. This chapter also covered the application, as well as giving a summary of recent UTM research.

References

1. Rumba, Rudolfs, and Agris Nikitenko. "The wild west of drones: A review on autonomous-UAV traffic-management." In *2020 International Conference on Unmanned Aircraft Systems (ICUAS)*, pp. 1317–1322. IEEE, 2020.
2. Yadav, Anshul, Salil Goel, Bharat Lohani, and Shubhanshi Singh. "A UAV Traffic Management System for India: Requirement and Preliminary Analysis." *Journal of the Indian Society of Remote Sensing* 49, no. 3 (2021): 515–525.
3. Ho, Florence, Artur Goncalves, Ana Salta, Marc Cavazza, Ruben Geraldes, and Helmut Prendinger. "Multi-agent path finding for UAV traffic management: Robotics track." (2019): 131–139.
4. Lin, Chin E., Tsung-Pao Chen, Pei-Chi Shao, Ying-Chih Lai, Tzung-Cheng Chen, and Yun-Chao Yeh. "Prototype Hierarchical UAS Traffic Management System in Taiwan." In *2019 Integrated Communications, Navigation and Surveillance Conference (ICNS)*, pp. 1–13. IEEE, 2019.
5. Ho, Florence, Rúben Geraldes, Artur Gonçalves, Bastien Rigault, Benjamin Sportich, Daisuke Kubo, Marc Cavazza, and Helmut Prendinger. "Decentralized multi-agent path finding for UAV traffic management." *IEEE Transactions on Intelligent Transportation Systems* (2020).
6. Kliushnikov, Ihor, Vyacheslav Kharchenko, Herman Fesenko, and Elena Zaitseva. "Multi-UAV Routing for Critical Iinfrastructure Monitoring

Considering Failures of UAVs: Reliability Models, Rerouting Algorithms, Industrial Case." In *2021 International Conference on Information and Digital Technologies (IDT)*, pp. 303–310. IEEE, 2021.

7. Wan, Shuo, Jiaxun Lu, Pingyi Fan, and Khaled B. Letaief. "Toward big data processing in IoT: Path planning and resource management of UAV base stations in mobile-edge computing system." *IEEE Internet of Things Journal* 7, no. 7 (2019): 5995–6009.
8. Smolyanskiy, Nikolai, Alexey Kamenev, Jeffrey Smith, and Stan Birchfield. "Toward low-flying autonomous MAV trail navigation using deep neural networks for environmental awareness." In *2017 IEEE/RSJ International Conference on Intelligent Robots and Systems (IROS)*, pp. 4241–4247. IEEE, 2017.
9. Fu, Zhangjie, Jingnan Yu, Guowu Xie, Yiming Chen, and Yuanhang Mao. "A heuristic evolutionary algorithm of UAV path planning." *Wireless Communications and Mobile Computing 2018* (2018).
10. Monwar, Momena, Omid Semiari, and Walid Saad. "Optimized path planning for inspection by unmanned aerial vehicles swarm with energy constraints." In *2018 IEEE Global Communications Conference (GLOBECOM)*, pp. 1–6. IEEE, 2018.
11. Zhan, Weiwei, Wei Wang, Nengcheng Chen, and Chao Wang. "Efficient UAV path planning with multiconstraints in a 3D large battlefield environment." *Mathematical Problems in Engineering 2014* (2014).
12. Kerns, Andrew J., Daniel P. Shepard, Jahshan A. Bhatti, and Todd E. Humphreys. "Unmanned aircraft capture and control via GPS spoofing." *Journal of Field Robotics* 31, no. 4 (2014): 617–636.
13. Seo, Seong-Hun, Byung-Hyun Lee, Sung-Hyuck Im, and Gyu-In Jee. "Effect of spoofing on unmanned aerial vehicle using counterfeited GPS signal." *Journal of Positioning, Navigation, and Timing* 4, no. 2 (2015): 57–65.
14. Kim, Alan, Brandon Wampler, James Goppert, Inseok Hwang, and Hal Aldridge. "Cyber attack vulnerabilities analysis for unmanned aerial vehicles." In *Infotech@ Aerospace 2012*, p. 2438. 2012.
15. Shashok, Nikolas. "Analysis of vulnerabilities in modern unmanned aircraft systems." Tuft University (2017): 1–10.
16. Sándor, Zsolt. "Challenges caused by the unmanned aerial vehicle in the air traffic management." *Periodica Polytechnica Transportation Engineering* 47, no. 2 (2019): 96–105.
17. Adhikari, Debesh, Eunjin Kim, and Hassan Reza. "A fuzzy adaptive differential evolution for multi-objective 3D UAV path optimization." In *2017 IEEE Congress on Evolutionary Computation (CEC)*, pp. 2258–2265. IEEE, 2017.
18. Piyush Agarwal, Sachin Sharma, and Priya Matta. "Components, Technologies, and Market of Road Traffic Management System in Global Scenarios: A Complete Study." In *2021 International Conference on Innovative Computing, Intelligent Communication and Smart Electrical Systems (ICSES)*, pp. 1–10. IEEE, 2021.

19. Anuj Singh, Tanmay Anand, Sachin Sharma, and Pankaj Singh. "IoT Based Weapons Detection System for Surveillance and Security Using YOLOV4." In *2021 6th International Conference on Communication and Electronics Systems (ICCES)*, pp. 488–493. IEEE, 2021.
20. Priya Kohli, Sachin Sharma, and Priya Matta. "Security Challenges, Applications and Vehicular Authentication Methods in VANET for Smart Traffic Management." In *2021 2nd International Conference on Intelligent Engineering and Management (ICIEM)*, pp. 327–332. IEEE, 2021.
21. Priya Kohli, Sachin Sharma, and Priya Matta. "Security of Cloud-Based Vehicular Ad-Hoc Communication Networks, Challenges and Solutions." In *2021 Sixth International Conference on Wireless Communications, Signal Processing and Networking (WiSPNET)*, pp. 283–287. IEEE, 2021.
22. Sakshi Painuly, Sachin Sharma, and Priya Matta. "Future Trends and Challenges in Next Generation Smart Application of 5G-IoT." In *2021 5th International Conference on Computing Methodologies and Communication (ICCMC)*, pp. 354–357. IEEE, 2021.
23. Amit Juyal, Sachin Sharma, and Priya Matta. "Deep Learning Methods for Object Detection in Autonomous Vehicles." In *2021 5th International Conference on Trends in Electronics and Informatics (ICOEI)*, pp. 751–755. IEEE, 2021.
24. Piyush Agarwal, Priya Matta, Sachin Sharma, "Analysis Based Traffic Flow Control Decision using IoT Sensors". *Materials Today: Proceedings* (2021).
25. Vartika Agarwal, and Sachin Sharma. "IoT based smart transport management system." In *International Conference on Advanced Informatics for Computing Research*, pp. 207–216. Springer, Singapore, 2020.
26. Vartika Agarwal, Sachin Sharma, and Piyush Agarwal. "IoT Based Smart Transport Management and Vehicle-to-Vehicle Communication System." In *Computer Networks, Big Data and IoT*, pp. 709–716. Springer, Singapore, 2021.
27. Sachin Sharma, Piyush Agarwal, and Seshadri Mohan. "Security Challenges and Future Aspects of Fifth Generation Vehicular Adhoc Networking (5G-VANET) in Connected Vehicles." In *2020 3rd International Conference on Intelligent Sustainable Systems (ICISS)*, pp. 1376–1380. IEEE, 2020.
28. Piyush Agarwal, Priya Matta, and Sachin Sharma. "Comparative Study of Emerging Internet-of-Things in Traffic Management System." In *2021 5th International Conference on Trends in Electronics and Informatics (ICOEI)*, pp. 422–428. IEEE, 2021.

11

Reliable Machine Learning-Based Detection for Cyber Security Attacks on Connected and Autonomous Vehicles

Ambika N.

Dept. of Computer Science and Applications, St. Francis College, Bangalore, India

Abstract

The previous system is driverless machines that take appropriate recommendations in all situations. They have transformed the conveyor system. The operations rely on sensed information and simulated knowledge to understand and reach conclusions concerning transportation operation, and to accommodate to improving circumstances. The method protection happens in three steps. Before the strikes occur, associated companies and businesses require determining class of advances that are liable to occur and their moderation practices. When the crimes occur, the operation should observe the complete CAV, and recognize initiatives. After raids follow, the method should respond to crimes appropriately and be capable to retrieve from seizures. The recommended UML-based CAV cyber protection structure is produced to determine the connections among each element and the construction in the CAV. Confined information has couple of sources. They contain device and software knowledge in the CAV frame. They have the running status info of appliance and programs. The Hardware section is the detector info obtained from the carrier enclosing by several CAV sensing devices. The program group has limited information assembled by the programs in CAVs. The surface course comprises information collected from other items. As each object has its own ID saved in its restricted knowledge, the visible details need this learning to support the testimony of the evidence source. After recognizing the transmitted ID, communications are classified into either individual stating that information is governed properly. The carriers or foundations need to transfer hidden knowledge. It can be obtained by distinct users and collected

Email: ambika.nagaraj76@gmail.com

Romil Rawat, A. Mary Sowjanya, Syed Imran Patel, Varshali Jaiswal, Imran Khan and Allam Balaram (eds.) Autonomous Vehicles Volume 1: Using Machine Intelligence, (203–216) © 2023 Scrivener Publishing LLC

in the safe location. The generalized category stocks information that everyone could obtain. The knowledge processing course has four essential data computation methodologies. The generating component accumulates knowledge from various origins. These datasets need to be formatted, organized and combined for computation. The Processing group computes the information, including cleaning or annotating the data for analysis. The Verification class includes components that make sure the data are secure, fulfilling the cyber security requirements of the CAV system. The recommended system has additional databases to take of the different kinds of assaults in the system. The reliability is increased by 7.32% and time taken to overcome an attack is decreased by 11.59% compared to the previous one.

Keywords: Connected and autonomous vehicle, cyber security, machine learning, reliability

11.1 Introduction

The vehicular conversation (Bhoi & Khilar, 2014) is an association among carriers. The principal purpose of stationing VANET is to decrease the level of collisions. It has impact on seated security. Motorists can ride easily in the metropolitan city. As vehicle communities progress, the percentage of disasters also grows. Hence the transports need to chat. VANET design essentially has carriers, Road Side component, and framework Field. Discussion accompanies using broadcast measures. The path component behaves like a routing component. It has high-range coverage than carriers reach. Wheels attaches with an On-Board subcomponent for exchanging data. It connects with a Global Positioning Scheme for understanding its location. It monitors other carriers. An automatic permit tag is set in the transport to identify them. The broadcast discovery service increases the simulated intensification of radioactivity methodologies. It supplies high reservoir energy. The certified authorization is in the planning that aids in implementing assistance.

Misconduct or irregularity discovery systems in vehicular arrangements are placed into three sections. The first section is a signature-based scheme. In this model, knowledge is assembled and then correlated with the database performance of specific offenses. The information corresponds with the bad action previously recorded. It is classified as an outbreak. This method cannot identify hidden strikes. The second class is related to specification-based practice. The plan meets these provisions. The protocol does not satisfy these requirements. It recognizes as a seizure. The third division assigns to an exception exposure operation. The machine training system builds a pattern of reliable and peculiarity exercise. It is the new response that examines this paradigm. This sort of arrangement can recognize unexplained crimes.

Computer knowledge (Qiu, Wu, Ding, Xu, & Feng, 2016) permits machines to discover private perspicacity. It uses iteratively reading from the information. It is explicitly programmed. It has transformed processor ability by providing education with massive datasets. It allows tools to modify, restructure and optimize procedures. The machine-learning comprises practice and examination. A prototype is determined based on the exercise knowledge in the training stage. The equipped design is employed to generate the forecast.

- Supervised knowledge - It is a device educational procedure (Cunningham, Cord, & Delany, 2008). It uses administered studying with identified data. The practice example has a ticket. The latest goal obtains the mapping from the data characteristic to the design. It is a credit forecast. It is classified into classification and regression methods. The distinction between the two jobs is that the tags are definite for class and statistical for regression.
- Unsupervised learning - The identification information hubs are the instructor in administered education. It judges the morality of the accomplished design in several conditions. It produces (Barlow, 1989) finding a description of the input examples without any labeling learning.
- Reinforcement learning - It is an operator (Sutton & Barto, 2018) that determines the optimal functioning through communicating with the atmosphere in a trial-and-error fashion. It intends to maximize bonuses from the atmosphere. The scene registers as a Markov judgment method. It includes activities and remunerates to a Markov means.
- Deep learning - It strives (Goodfellow, Bengio, & Courville, 2016) to acquire information descriptions. It can be built-in managed, unsupervised, and support knowledge. It has made notable progressions in different motor knowledge responsibilities.

The contribution (He, Meng, Qu, & Xi, 2020) focuses on the secrecy and protection prospects in Independent carriers. The driverless machines (Gnatzig, Schuller, & Lienkamp, 2012) (McBride, 2016) take appropriate recommendations in all situations. They have transformed the conveyor system. The operations rely on sensed information and simulated knowledge to understand and reach conclusions concerning transportation operation, and to accommodate to improving circumstances. The method protection happens in three steps. Before the strikes occur, associated companies and businesses require determining class of advances that are liable to occur and their moderation practices. When the crimes appear, the operation should observe the complete CAV, and recognize initiatives. After raids follow, the method should respond to crimes appropriately and be capable to retrieve from seizures. The recommended UML-based CAV cyber protection structure is produced to determine the connections among each element and the construction in the CAV. Confined information has a couple of sources. They contain device and software knowledge in the CAV frame. They have the running status information of appliance and programs. The Hardware section is the detector information obtained from the carrier enclosing by several CAV sensing devices. The program group has limited information assembled by the programs in CAVs. The surface course comprises information collected from other items. As each object has its own ID saved in its restricted knowledge, the visible details need this learning to support the testimony of the evidence source. After recognizing the transmitted ID, communications are classified into either individual stating that information is governed properly. The carriers or foundations need to transfer hidden knowledge. It can be obtained by distinct users and collected in the safe location. The generalized category stocks information that everyone could obtain. The knowledge processing course has four essential data computation methodologies. The generating component accumulates knowledge from various origins. These datasets need to be formatted, organized and combined for computation. The Processing group computes the information, including cleaning or annotating the data for analysis. The Verification class includes components that make sure the data are secure, fulfilling the cyber security (Ambika N., 2021) (Ambika N., 2020) requirements of the CAV system. The recommended system has additional databases to take of the different kinds of assaults in the system. The reliability is increased by 7.32% and time taken to overcome an attack is decreased by 11.59% compared to the previous one.

The chapter is divided into five sections. The literature survey follows the introduction in section two. The proposed architecture is narrated in segment three. The experimental setup is listed in section four. The analysis is explained in segment five. The conclusion is presented in section six.

11.2 Literature Survey

The section contains the contributions made by various authors in the domain. The contribution (He, Meng, Qu, & Xi, 2020) focuses on the secrecy and protection prospects in Independent carriers. The driverless machines (Gnatzig, Schuller, & Lienkamp, 2012) (McBride, 2016) take appropriate recommendations in all situations. They have transformed the conveyor system. The operations rely on sensed information and simulated knowledge to understand and compose conclusions concerning transportation operation, and to accommodate to improving circumstances. The method protection happens in three steps. Before the strikes occur, associated companies and businesses require determining class of advances that are liable to occur and their moderation practices. When the crimes appear, the operation should observe the complete CAV, and recognize initiatives. After raids follow, the method should respond to crimes appropriately and be capable to retrieve from seizures. The recommended UML-based CAV cyber protection structure is produced to determine the connections among each element and the construction in the CAV. Confined information has a couple of sources. They contain device and software knowledge in the CAV frame. They have the running status information of appliance and programs. The Hardware section is the detector information obtained from the carrier enclosing by several CAV sensing devices. The program group has limited information assembled by the programs in CAVs. The surface course comprises information collected from other items. As each object has its own ID saved in its restricted knowledge, the visible details need this learning to support the testimony of the evidence source. After recognizing the transmitted ID, communications are classified into either individual stating that information is governed properly. The carriers or foundations need to transfer hidden knowledge. It can be obtained by distinct users and collected in the safe location. The generalized category stocks information that everyone could obtain. The knowledge processing course has four essential data computation methodologies. The generating component accumulates knowledge from various origins. These datasets need to be formatted, organized and combined for computation. The Processing group computes the information, including cleaning or annotating the data for analysis. The Verification class includes components that make sure the data are secure, fulfilling the cyber security requirements of the CAV system.

The transports (Gyawali & Qian, Misbehavior detection using machine learning in vehicular communication networks, 2019) can determine the number of carriers and the medium velocity of wheels within its transmission lattice. Based on mediocre pace and weight, each wheel can measure the amount of movement. The transport telecast estimates the value of this course to its acquaintance. The transportation may get this price of progress from various transports within a specific period. Based on the accepted assessment of issues from all next-door-neighbor, they estimate the medium amount of course. Based on the assessed progress and earned rate of movement, each carrier can receive a new characteristic signature variation inflow. It is the diversity of medium and earned amount of progress. Each moment it accepts an observant signal from its bystander. It gives the message with advice to its limited discovery arrangement. Each bounded disclosure method uses either traditional datasets or datasets produced through simulations to check the authenticity of a flag. In the site falsification crime, the enemy switches the location learning in the signal. In a permanent status assault, the attacker broadcasts determined area knowledge whereas, in a constant offset mugging, the criminal dispatches solidified offset combined to their original condition. The system indicates an intervention based on the different characteristics produced from the accepted flag. Each moment a transport gets a signal from different wheels, it examines its learning with the prior one and limited data of its radar. Based on this erudition, each carrier can create several peculiarities.

It is a computer learning-based discovery operation (Gyawali, Qian, & Hu, Machine learning and reputation based misbehavior detection in vehicular communication networks, 2020). It detects multiple misbehaviors in communications. The reliability record identifies the integrity of the wheels. Carriers obtain enough new status passes through entrance spots. Each transportation can confirm the honesty of a destination transport by judging the reliability count. The information from these devices goes to the motor learning-based system. The framework is on each wheel. It transmits the effect of discovery as feedback to the legitimate local authority. It employs Dempster Shafer ideas and faith records of the transportation for feedback sequence. The outcome achieves after the feedback sequence updates the reliability score of the target vehicle. A carrier can ask for the newest reliability certification when it runs near the admittance locality. The gathering wheels measure the discrepancy within the prevailing time and credit record production time. It matches with an outset content. The transport is accepted as reliable if it passes the test. Then the communication is transmitted to the misconduct discovery arrangement for additional affirmation. Both wheels convey the credit application to the certificate authority. It creates reliability guarantees for both means of transport and communicates it to them. Once wheels get reliability guarantees, they can start announcing the information.

The suggestion (Sliwa & Wietfeld, 2019) is a data-driven strategy. It is to simulate the system management of a detailed real-world situation. Administered training is employed to obtain a deterministic forecast design. It permits the anticipation of the performance of the analyzed end-to-end symbol based on the evidence of the presented circumstances. The pattern flaws require exercising into record within the simulations. A pragmatic estimation is obtained within the DDNS by sampling from the failure frequency of the real-life determinations. Another pattern is employed to modify the foresight pattern. It alters from the deterministic to the probabilistic area. Prognosticating end-to-end execution symbols is a regression assignment.

The suggested approach (Bhatti, *et al.*, 2021) forms the batches using simulated knowledge. The procedure builds groups using the computer education method. It picks head based on engagement and availability from carriers list. The work assesses the signal-to-noise proportion of all cognitive vehicles. It cannot cooperate for the sensing if the estimated SNR is less than a predefined outset. The device determines the batches of the picked transport. It chooses the head for them using a device training procedure. The method examines all vehicles as possible leaders. It interchanges communications among vehicles until a set of leads emerges. The practice attempts on knowledge. Each vehicle switches the engagement and availability with all other transports. The repetitions stop when engaged heads stay constant for a duo of redundancies.

The recommendation (Yang, *et al.*, 2020) has a preparation method and testimony manner. The situation description applies to administer training in motor knowledge. Each information vector in the practice data has a distinguished production. The procedure adapts the design so that the yield of the pattern resembles the anticipated outcomes. The knowledge is the framework of the building has carrier interference paradigm. A list of the systematic flow of features and the similar situations are exercise knowledge. The arrangement uses a back-propagation Fuzzy System. The learning used to teach the system comes from course computations. An assemblage of flow estimation performs course measures under situations with or without transport barriers. The antenna and recipient are on two wheels. The focus apparatus is a vector beacon communications PXIe-5673 of National Instruments. It gets the baseband computation sign produced by the machine and arrives at the core recurrence of 5.9 GHz. It has the vector sign examiner. It down-convert wireless regularity (RF) indicates to baseband and then stores them in evidence record.

11.3 Proposed Architecture

The contribution is enhanced structure from the previous architecture (He, Meng, Qu, & Xi, 2020). Some additional components are added to the past system to make it better. The system has additional databases:

1. Database of different kinds of attacks (liable in the region) as in (He, Meng, Qu, & Xi, 2020)
2. Database of attacks similar to the (1) in other regions
3. Database of hybrid-based attacks (along with its summary) found in other regions.

The cloud (Nagaraj, 2021) is used to store the databases. The databases contain the feed of the attack summary, its behavior in the environment, time it may sustain, different phase of the attacks, suggestions taken to overcome the attack. The databases are used accordingly. After overcoming the attack, the analysis result along with other details is updated in the respective database.

The working of the system is similar to the previous recommendation (He, Meng, Qu, & Xi, 2020). The driverless machines (Gnatzig, Schuller, & Lienkamp, 2012) (McBride, 2016) take appropriate recommendations in all situations. They have transformed the conveyor system. The operations rely on sensed information and simulated knowledge to understand and compose conclusions concerning transportation operation, and to accommodate to improving circumstances. The method protection happens in three steps. Before the strikes occur, associated companies and businesses require determining class of advances that are liable to occur and their moderation practices. When the crimes appear, the operation should observe the complete CAV, and recognize initiatives. After raids follow, the method should response to crimes appropriately and be capable to retrieve from seizures. The recommended UML-based CAV cyber protection structure is produced to determine the connections among each element and the construction in the CAV. Confined information has a couple of sources. They contain device and software knowledge in the CAV frame. They have the running status information of appliance and programs. The Hardware section is the detector information obtained from the carrier enclosing by several CAV sensing devices. The program group has limited information assembled by the programs in CAVs. The surface course comprises information collected from other items. As each object has its own ID saved in its restricted knowledge, the visible details need this learning to support the testimony of the evidence

Table 11.1 AI algorithm used to select the appropriate database.

Step 1: By default the first database is used.
Step 2: Enter the analyzed parameters (behavior of the assault).
Step 3: Based on the outcome of Step 2, the respective database is used.
Step 4: The Phase of the present attack is mapped with the database phases stored.
Step 5: The measure to be adopted is suggested to the system.

source. After recognizing the transmitted ID, communications are classified into either individual stating that information is governed properly. The carriers or foundations need to transfer hidden knowledge. It can be obtained by distinct users and collected in the safe location. The generalized category stocks information that everyone could obtain. The knowledge processing course has four essential data computation methodologies. The generating component accumulates knowledge from various origins. These datasets need to be formatted, organized and combined for computation. The Processing group computes the information, including cleaning or annotating the data for analysis. The Verification class includes components that make sure the data are secure, fulfilling the cyber security requirements of the CAV system. Table 11.1 shows about the AI algorithm used to select the appropriate database.

11.4 Experimental Results

The work is simulated in NS2. Table 11.2 is the set of parameters used in the system.

11.5 Analysis of the Proposal

- *Increasing reliability* – The previous system (He, Meng, Qu, & Xi, 2020) uses a subset of database where the listing of the attacks is limited to the attacks happening in the region. As the technology is advancing, the attacks are mutated and also new ones are introduced. To tackle these kinds of attacks, new databases accommodating the scalability of attacks are to be analyzed. The suggestion takes care of the

Table 11.2 Parameters used in the proposal.

Parameters used	**Description**
Dimension of the environment considered	200 m * 200m
Size of database 1	10 MB
Size of database 2	12 MB
Size of database 3	6 MB
Number of vehicles used	3
Speed of communication	200 km/hr
Bandwidth of the communication media	75 MHz
Type of attacks considered	Type1 (database1) -2; type2 (database2) -1; type3 (database3) -2
Length of packets communicated at a instance	5.9 GHz
Time of simulation	60

same. The behavior of the assault can be analyzed better and appropriate action can be taken. An AI algorithm is used to make a choice between the databases to be utilized at that instance. The recommendation proves to be more reliable by 7.32% compared to the past work. The same is represented in Figure 11.1.

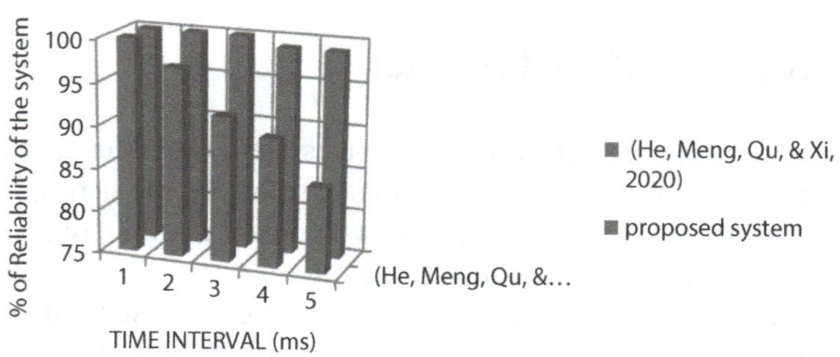

Figure 11.1 Representation of the comparison of both work w.r.t reliability.

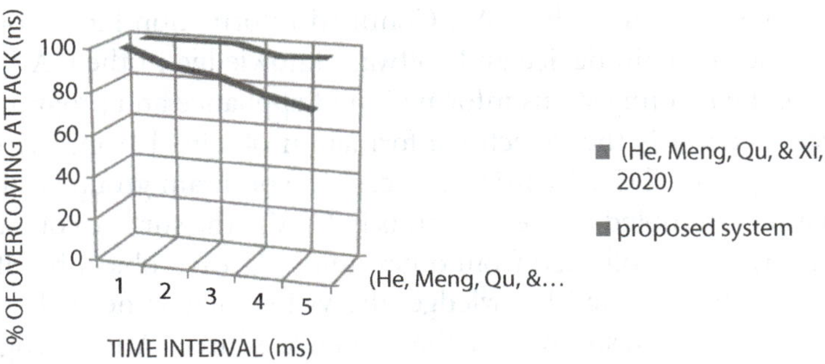

Figure 11.2 Graphical representation of the speed to overcome the attack.

- Increasing speed of recovery – The past structure uses a single database containing the assault details of the previous attacks occurred. The suggestion considers other possibility to take care of various kinds of attacks. It also contains the detailing of the overcoming measures compared to the previous system (He, Meng, Qu, & Xi, 2020). Figure 11.2 is the graphical representation of the time taken to overcome the attack. The time taken to overcome an assault is decreased by 11.59% higher than the previous one.

11.6 Conclusion

Vehicular communication is the exchanging of data between the vehicles to avoid accidents and avoid traffic congestion. These devices receive or transmit information that enables safety warnings and traffic information. These systems are capable for short-range communication. The previous system is driverless machines that take appropriate recommendations in all situations. They have transformed the conveyor system. The operations rely on sensed information and simulated knowledge to understand and compose conclusions concerning transportation operation, and to accommodate to improving circumstances. The method protection happens in three steps. Before the strikes occur, associated companies and businesses require determining class of advances that are liable to occur and their moderation practices. When the crimes appear, the operation should observe the complete CAV, and recognize initiatives. After raids follow, the method should respond to crimes appropriately and be capable to retrieve from seizures. The recommended UML-based CAV cyber protection structure is produced to determine the connections among each element and the construction in the CAV. Confined information has a couple of sources. They contain device and software knowledge in the CAV frame. They have the running status information of appliance and programs. The Hardware section is the detector information obtained from the carrier enclosing by several CAV sensing devices. The program group has limited information assembled by the programs in CAVs. The surface course comprises information collected from other items. As each object has its own ID saved in its restricted knowledge, the visible details need this learning to support the testimony of the evidence source. After recognizing the transmitted ID, communications are classified into either individual stating that information is governed properly. The carriers or foundations need to transfer hidden knowledge. It can be obtained by distinct users and collected in the safe location. The generalized category stocks information that everyone could obtain. The knowledge processing course has four essential data computation methodologies. The generating component accumulates knowledge from various origins. These datasets need to be formatted, organized and combined for computation. The Processing group computes the information, including cleaning or annotating the data for analysis. The Verification class includes components that make sure the data are secure, fulfilling the cyber security requirements of the CAV system. The recommended system has additional databases to take of the different kinds of assaults in the system. The reliability is increased by 7.32% and time taken to overcome an attack is decreased by 11.59% compared to the previous one.

References

Ambika, N. (2021). An Improved Solution to Tackle Cyber Attacks. In F. Al-Turjman, & N. Gowthaman, *Advanced Controllers for Smart Cities: An Industry 4.0 Perspective* (pp. 15–23). Switzerland AG: SPRINGER NATURE.

Ambika, N. (2020). Improved Methodology to Detect Advanced Persistent Threat Attacks. In N. K. Chaubey, & B. B. Prajapati, *Quantum Cryptography and the Future of Cyber Security* (pp. 184–202). USA: IGI Global.

Barlow, H. B. (1989). Unsupervised learning. *Neural computation, 1* (3), 295–311.

Bhatti, S, D. M., Rehman, Y., Rajput, P. S., Ahmed, S., Kumar, P., *et al.* (2021). Machine learning based cluster formation in vehicular communication. *Telecommunication Systems*, 1–9.

Bhoi, S. K., & Khilar, P. M. (2014). Vehicular communication: a survey. *IET Networks, 3* (3), 204–217.

Cunningham, P., Cord, M., & Delany, S. J. (2008). Supervised learning. In C. P. Cord M., *Machine learning techniques for multimedia* (pp. 21–49). Berlin, Heidelberg: Springer.

Gnatzig, S., Schuller, F., & Lienkamp, M. (2012). Human-machine interaction as key technology for driverless driving-A trajectory-based shared autonomy control approach. *21st IEEE International Symposium on Robot and Human Interactive Communication* (pp. 913–918). Paris, France: IEEE.

Goodfellow, I., Bengio, Y., & Courville, A. (2016). *Deep learning.* Massachusetts Avenue: MIT Press.

Gyawali, S., & Qian, Y. (2019). Misbehavior detection using machine learning in vehicular communication networks. *IEEE International Conference on Communications (ICC)* (pp. 1–6). Shanghai, China: IEEE.

Gyawali, S., Qian, Y., & Hu, R. Q. (2020). Machine learning and reputation based misbehavior detection in vehicular communication networks. *IEEE Transactions on Vehicular Technology, 69* (8), 8871–8885.

He, Q., Meng, X., Qu, R., & Xi, R. (2020). Machine Learning-Based Detection for Cyber Security Attacks on Connected and Autonomous Vehicles. *Mathematics, 8* (8), 1–19.

McBride, N. (2016). The ethics of driverless cars. *Acm Sigcas Computers and Society, 45* (3), 179–184.

Nagaraj, A. (2021). *Introduction to Sensors in IoT and Cloud Computing Applications.* UAE: Bentham Science Publishers.

Qiu, J., Wu, Q., Ding, G., Xu, Y., & Feng, S. (2016). A survey of machine learning for big data processing. *EURASIP Journal on Advances in Signal Processing, 2016* (1), 1–16.

Sliwa, B., & Wietfeld, C. (2019). Data-driven network simulation for performance analysis of anticipatory vehicular communication systems. *IEEE Access, 7,* 172638–172653.

Sutton, R. S., & Barto, A. G. (2018). *Reinforcement learning: An introduction.* Massachusetts Avenue: MIT Press.

Yang, M., Ai, B., He, R., Huang, C., Li, J., Ma, Z., *et al.* (2020). Identification of vehicle obstruction scenario based on machine learning in vehicle-to-vehicle communications. *IEEE 91st Vehicular Technology Conference (VTC2020-Spring)* (pp. 1–5). Antwerp, Belgium: IEEE.

12

Multitask Learning for Security and Privacy in IoV (Internet of Vehicles)

Malik Mustafa[1], Ahmed Mateen Buttar[2], Guna Sekhar Sajja[3], Sanjeev Gour[4], Mohd Naved[5] and P. William[6]*

[1]Faculty of Business and Management Studies, Center for Foundation Studies, Gulf College, Muscat, Oman
[2]Department of Computer Science, University of Agriculture, Faisalabad, Pakistan
[3]Department of Information Technology, University of the Cumberlands, Williamsburg, Kentucky, United States
[4]Department of Computer Science, Career College, Bhopal, Madhya Pradesh, India
[5]Department of Business Analytics, Jagannath University, Delhi-NCR, India
[6]Department of Information Technology, Sanjivani College of Engineering, SPPU, Pune, India

Abstract
An AI system simulates human intellect in order to build a computer capable of addressing real-world issues. Machine learning is a field in artificial intelligence and a proven prediction tool in many real-world applications. Real-world issues in society have been successfully addressed using these methods. The Internet of Things (IoT) has recently evolved into the Internet of Vehicles (IoV), where vehicles in a setting connect with many facets of society through the IoT. Fully autonomous cars and semi-autonomous vehicles coexist on the IoV. Both exterior and interior surroundings may be reached by these vehicles in the IoT. The cars communicate with other vehicles, the road infrastructure, sensors, personal gadgets, pedestrians, and residences. Over the next several years, each car is likely to have over 200 sensors installed. It will generate an enormous amount of data. This data is required to be collected and processed effectively. Also in IoV, security of data is crucial. This article contains a comprehensive study of application of artificial intelligence and machine learning on IoV network data, to predict all potential threats to secure IoV from various attacks and threats.

**Corresponding author*: william160891@gmail.com

Romil Rawat, A. Mary Sowjanya, Syed Imran Patel, Varshali Jaiswal, Imran Khan and Allam Balaram (eds.) Autonomous Vehicles Volume 1: Using Machine Intelligence, (217–234) © 2023 Scrivener Publishing LLC

***Keywords*:** IoV, IoT, attacks, threats, artificial intelligence, prediction

12.1 Introduction

The Internet of Things (IoT) has recently evolved into the Internet of Vehicles (IoV), which connects vehicles in a setting to many different aspects of society. On the IoV, there are both fully autonomous and semi-autonomous vehicles. In the IoT, these vehicles can access both their outer and internal environments. They communicate with other vehicles, road infrastructure, sensors and personal gadgets, pedestrians, and homes. Sensors are expected to be deployed in every vehicle within the next several years. A massive amount of data will be generated. The collection and processing of this data is essential. Data security is critical in the Internet of Vehicles as well.

The IoT is a hot topic right now. Everything in the real world is being affected by it. It has countless uses and is enhancing people's lives. Because of IoT technology, concepts like smart city, smart home, smart healthcare, smart agriculture, smart farming, smart logistics, and smart production were all viable. This is shown below in Figure 12.1.

There are several benefits to the Internet of Things, such as enhancing efficiency, reducing costs, maximizing energy, and ensuring forecasts. Security dangers are developing as more technologies and data processing become more diversified. As a result of security and privacy issues, however, IoT has had a difficult time taking off. Security and privacy challenges are shown in Figure 12.2.

In computer science, a network is a collection of computers or devices that are linked together and may access the same resources. Before networks were created, there were freestanding computers that took part in stand-alone computing. In the future, computer technology will undergo a number of changes. It is important to note that there are three main forms of computing: centralized, client/server, and distributed. Grid computing and cloud computing are two instances of distributed computing. The IoT is made feasible through distributed computing.

For example, it is the technology that includes technologies from several domains such as computer science, electronic engineering and so on. The IoT relies on a wide range of disciplines and standards from a variety of suppliers to enable its ability to connect physical and digital objects. Many real-world IoT applications may be found in this setting. Precision agriculture, for example, uses a variety of technology to determine the inputs crops need and when they need them. It is possible to boost productivity and reduce waste by making the correct decisions at the appropriate time.

Figure 12.1 Internet of Things in real-life applications.

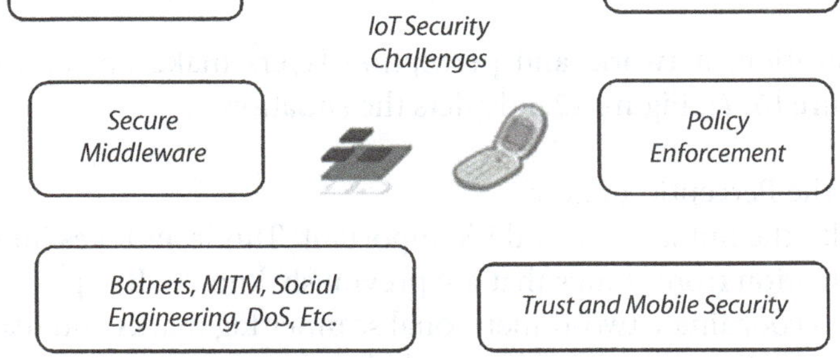

Figure 12.2 Security and privacy challenges in IoV.

Smart homes, smart healthcare systems, smart cities, intelligent transportation systems, and inventory tracking are just a few examples of other IoT applications [1, 2]. For example, a wide range of applications might benefit from IoT's incorporation. The Internet of Things has the potential to enhance device connection. Thus, M2M communication can take place without the need for human intervention. As a result of the IoT, machine inefficiencies will be reduced. Data is collected through sensors in IoT systems.

Using data that has been gathered and evaluated, it is possible to make well-informed decisions. Automation and control are made easier by the IoT [3]. Easy access to devices and their data allows for more elegant automation and control. The IoT helps to monitor applications. It can help with successful monitoring since it predicts incidents. Other uses include assessing water distribution, supply chain enterprises, energy distribution and supply amount.

In real-world systems, M2M will reduce response time and boost efficiency. Besides this, the IoT can help people in their daily life. Everything becomes more productive, efficient, and forward-thinking as a result of this technology. The Internet of Things can connect the digital and physical worlds. Unlike a human, for example, a cell phone is digital in nature. The IoT properly connects these two.

Technology such as Radio Frequency Identification (RFID), sensing, and other IoT gadgets are used to achieve this goal. If you're looking for a solution that provides on-demand storage and processing, cloud computing could be a better option. Consequently, cloud computing and the IoT are necessary for real-world applications. The use of distributed programming frameworks like Hadoop for data storage and administration is also enabled by cloud computing.

12.2 IoT Architecture [5]

The application, network, and perception layers make up the three-tier architecture [6, 7]. Figure 12.3 depicts the situation.

- The Perception Layer
  It's the initial layer, and it's important. This layer takes information from things that are previously known. It is possible to coordinate two-dimensional scanner tag marks and readers using various sensors, including cameras, terminals and remote sensors. QR codes and sensors are integrated onto

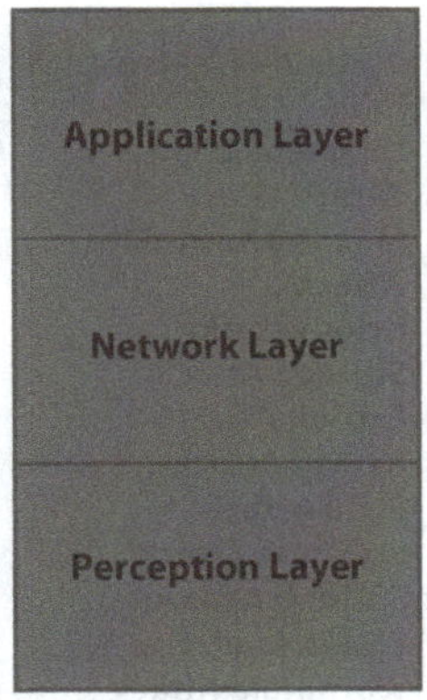

Figure 12.3 Three layer architecture.

RFID labels. This layer excels in object detection and data collection.

- The Network Layer
  The network layer is also referred to as the "transmission" layer. IoT devices are protected against unauthorised access to data from the layer of perception in this layer. This layer manages firewalls, hubs, and switches for networking needs.
- The Application Layer
  It sits between the market and the middleware layers. The platform layer provides consumers with app-related resources. This sheet can accommodate a wide range of applications, including smart health, smart home, smart environment, smart automobile, and so on.

12.3 Taxonomy of Various Security Attacks in Internet of Things [5]

Attacks on the IoT are categorised in Figure 12.4. Based on a three-layer structure, this taxonomy was created. Attacks on the perception, network, and application layers are included.

12.3.1 Perception Layer Attacks

Attackers may easily get access to tags because RFID systems lack robust authentication methods [8–10]. Data may be misused by attackers. Once

Application Layer Attack	Network Layer Attacks	Perception Layer Attacks
•Cross Site Scripting,	•Denial of Service Attack,	•Unauthorised Access To Tags
•Malicious Code Attack	•Man In The Middle Attack,	•Tag Cloning
•Buffer Overflow	•Ddos Attack,	•Eavesdropping
•Phishing Attack	•Storage Attack,	•Tracking Attack
•Authentication And Authorisation	•Exploit Attack,	•RF Jamming
•Sensitive Data Permission Or Manipulation	•Replay Attacks,	•Node Jamming
•Web Browser Attack	•Sybil Attack,	•Spoofing Attack
•Sql Injection Attack	•Sinkhole Attack,	•Node Capture Attack
	•Sniffing Attack,	•Malicious Node Attack
	•Traffic Analysis,	•Timing Attack
	•Wormhole Attack	•Side Channel Attack
	•Routing Information	

Figure 12.4 Taxonomy of various securities related attacks in IoT based on architecture layers [5].

an intruder has obtained access to a wireless sensor network, he or she may initiate an attack or make use of the network at no cost.

Cloned RFID tags are an effective attack method. Attackers may get access to the operating environment by using reverse engineering or by outright breaking into the system [8–11]. Previous studies [8] have found compromises because RFID scanners cannot tell one tag from another.

Wireless connections allow the intruder to monitor the sensor's device and node effectively [8, 11]. Using an antenna in an RFID system, a reader and a tag can communicate with each other [12]. Antennas may be used by unauthorised parties to intercept data sent between a reader and a tag [13].

The system uses RF signals to keep the legal tag from being read [8–14]. Intruders often utilise RFID tags that restrict readers from interacting with any tags within their range [13]. Depending on the amount of knowledge, this form of attack might do significant damage to the systems used to collect data.

For the same level of authorization or operation, all tags can be disguised as genuine. As a result, the reader will be misled into thinking it has the same permissions as the authorised tag. As a legitimate tag, an intruder must have access and a thorough comprehension of procedures [13]. Once an attacker has gained access to a contact channel, they must also have access to a similar contact channel. During transmission, packets might be lost due to spoofing attempts [15]. Due to nodes being forced to repeat data transfers, a massive attack like this could lead to an increase in network traffic of significant proportion.

The battery's power limits the device and the visual layer's node. Once it is no longer in use, the system must be placed into "sleep mode." These attacks try to disrupt the system by constantly supplying control data to the system and keeping the node active [8, 27].

12.3.2 Network Layer Attacks

Network DoS attacks create a significant amount of bandwidth [8, 15, 16] as a consequence. So users may have difficulty using the network or its resources. Data may still be stolen even if it is encrypted [8]. To attack several targets at the same time, a DDoS assault might employ a multi-computer attachment platform.

This allows the target node to carry out an action twice, eliminating redundancy, in Sybil attacks [8, 15, 17]. Due to the attacker's various identities, the victim's node may transfer information via a hacked node that results in a longer route on the wireless network (WSN).

Attackers use the included node to harvest data from the surrounding nodes during a sinkhole attack. Already, in [8], devices and data have been interfered with. To steal network traffic and sensor data, an attacker needs just one WSN rogue node. Assailants would utilise sniffer devices and software in an attempt to gain access to the network and steal important data.

Detecting communication patterns and load is based on the number and size of data packets [10, 19]. More packets can be analysed to acquire more relevant information. This technique might potentially be extended to incorporate encrypted packets [4, 18], as well. It is also possible to examine the pattern of transmission. There are three approaches to monitor WSN traffic [21–23]. To conclude, an intruder can determine the network's current state of affairs. An intruder [24–26] must be present in order to discover wireless connection points (APs) [28]. An attacker may [29] also be able to determine the format in which data is being sent.

Both sides are being listened in on by assailants. The constant exchange of received messages between communication partners makes communication services more difficult. The reader-to-tag correlation of RFID technology is likewise susceptible to this attack type. Both the reader and tag databases, as well as the backend processing resources, are being depleted by the assault. As a bonus, radio signal transmission can be used to gain access to the reader.

A man-in-the-middle attack places the attacker in the middle of two targets at the same time. Attackers connect two victims to a lawful node hidden by the perpetrator [8–10, 20]. A total of two nodes and two victims have been identified.

12.3.3 Application Layer Attacks

Network DoS attacks create a significant amount of bandwidth [8, 15, 16] as a consequence. So users may have difficulty using the network or its resources. Data may still be stolen even if it is encrypted [8]. To attack several targets at the same time, a DDoS assault might employ a multi-computer attachment platform.

This allows the target node to carry out an action twice, eliminating redundancy, in Sybil attacks [8, 15, 17]. Due to the attacker's various identities, the victim's node may transfer information via a hacked node that results in a longer route on the wireless network (WSN).

Attackers use the included node to harvest data from the surrounding nodes during a sinkhole attack. Already, in [8], devices and data have been interfered with. To steal network traffic and sensor data, an attacker needs just one WSN rogue node. Assailants would utilise sniffer devices and software in an attempt to gain access to the network and steal important data.

Detecting communication patterns and load is based on the number and size of data packets [10, 19]. More packets can be analysed to acquire more relevant information. This technique might potentially be extended to incorporate encrypted packets, as well. It is also possible to examine the pattern of transmission. There are three approaches to monitor WSN traffic. To conclude, an intruder can determine the network's current state of affairs. An intruder must be present in order to discover wireless connection points (APs). An attacker may also be able to determine the format in which data is being sent.

Both sides are being listened in on by assailants. The constant exchange of received messages between communication partners makes communication services more difficult. The reader-to-tag correlation of RFID technology is likewise susceptible to this attack type. Both the reader and tag databases, as well as the backend processing resources, are being depleted by the assault. As a bonus, radio signal transmission can be used to gain access to the reader.

A man-in-the-middle attack places the attacker in the middle of two targets at the same time. Attackers connect two victims to a lawful node hidden by the perpetrator [8–10, 20]. A total of two nodes and two victims have been identified.

12.4 Machine Learning Algorithms for Security and Privacy in IoV

To put it another way, machine learning (ML) is a new field of data mining in which an algorithm may learn and improve without needing to be explicitly coded [30]. Unsupervised and supervised learning are two of the most frequent methods of machine learning. There are two types of inference: classification and regression. For classification, the training data must be labelled; for regression, the training data must be unlabeled (clustering).

To classify anything, you first need to identify the characteristics that make it distinctive, which are known as target attributes or labels. Bayesian classifiers are used in a wide range of applications [31], as are decision trees and support vector machines. They are also employed in artificial neural networks and closest neighbour classifiers. Each one will be briefly discussed. Each technique relies on learning algorithms as its foundation.

When faced with categorization challenges, basic and easy decision trees might help. A decision tree categorises events based on their attribute values. Classifications are represented by nodes and branches in the decision tree, and each branch represents a potential value for a node. Root nodes classify instances, whereas feature values are used to rank them in the hierarchy.

In certain cases, it may be difficult to precisely characterise input attributes. However, class variables are non-deterministic even if the input attributes set values match those in the training dataset. This is due to the processing of the data not taking into account some noise and confusion. Cardiovascular disease may be detected in the course of daily life, for example.

Some individuals who consume a healthy diet and exercise frequently are nonetheless at risk of developing heart disease because of other factors such as smoking, alcohol intake, and maybe a genetic predisposition. In some cases, the well-known features of heart disease are insufficient to provide trustworthy information. Attributes are used by the Bayesian classifier to deduce the class label in these situations [32].

Biological neural networks are utilised to develop animal brains. The fact that it is composed of connected nodes and directed links makes ANN a connectionist system. Node to node communication is accomplished via the use of connections, each of which carries a certain amount of weight. Each node in the network must first process signals before they can be sent on to the next.

In most artificial neural network implementations, a non-linear function is applied to the output of each neuron depending on the sum of all its inputs. When learning proceeds, the signal intensity grows or falls depending on the weight of artificial neurons and their connections [33].

In ML classification, there are two techniques to construct a learning model. Learning begins immediately when the training set is made available to these models. As long as the properties of the test case completely match any of the training examples, a separate model may identify it. These students are described as "lazy students" [34].

This categorization has d features and each sample is treated as a single data point in the space of all the other samples. It is established how far apart the test example is from the rest of the data. All the points in the vicinity of a data point are counted as neighbours.

According to its neighbours' classifications, the data point is classified. The class label that is most commonly used by other data points is assigned to data points with several classed neighbours. To accomplish the work, the exact value of k's closest neighbours must be established. Noise in training data may lead to misclassification if k is set to a small number. When the number of closest neighbours exceeds k, it may contain data points that are outside of the test attribute's immediate area. Inaccurate categorisation might result from this.

In the beginning, random forest is a machine learning technique that builds a forest of judgments using multiple decision trees built from random vector input. This approach may be used to solve issues with classification and regression. If the number of trees in the random forest is increased, the random forest's forecasts will have a better degree of accuracy. It is vital to understand that decision trees are not the same as a forest.

The primary distinction between decision trees and random forests is the random splitting of feature nodes in order to find the root node. Random forest classification is often used because of its benefits. Its ability to be used for both classification and regression analysis is one of its many benefits. Additionally, this technique avoids the issue of overfitting if a large number of trees are accessible. It is also possible to express categorical data using random forest classification, which can cope with missing data.

From retail to healthcare to banking and the stock market, random forest classifiers may be used in a variety of industries. Random classifiers are used in banking to differentiate between authentic customers and fraudsters. In medicine, random forest is used to discover the best combination of medications and to diagnose sickness using the history of a patient's medical conditions observing a stock's behaviour and then determining its profit or loss is done using random forest classifier in the financial sector. Predicting what goods consumers will purchase on the basis of their preferences may be done using random forest in e-commerce settings.

The Support Vector Machine (SVM) is the best tool for categorisation. As a consequence, everyone in the company is buzzing with anticipation. Using an SVM model, it is possible to clearly discriminate between instances of various categories in vector space. New samples are assigned a label depending on which side of the gap they fall into [35] when they are mapped into the chosen vector space. Using the kernel approach, an SVM is capable of non-linear classification.

12.5 A Machine Learning-Based Learning Analytics Methodology for Security and Privacy in Internet of Vehicles

12.5.1 Methodology

Three machine learning algorithms make up the methodology. They include ID3, SVM, and artificial neural network (ANN).

As the first such decision tree-based system to develop, J. Ross Quinlan devised the ID-3 algorithm (Iterative otomiser-3). Entropy and information gain are considered in the computations of this method. Each dataset iteration includes a calculation of the functional attribute entropy values. It is taken into consideration when separating datasets into subgroups based on which characteristics have lower error rates (entropy) (information gain). It is necessary to repeat the process on each subset in order to verify that the data has been correctly classified. A decision tree may be constructed by starting at a non-terminal node and working your way down to the terminal nodes at the branch's termination point. The split property specifies nonterminal nodes, while terminal nodes are class labels for a certain class. To better identify and predict cardiac sickness in its early stages, researchers used ID-3-based decision trees [36]. Building a prediction model based on a large health data collection involves using CART and ID-3 decision tree techniques. The findings are verified using a 10-fold cross validation procedure. Decision tree classification methods may be used to build an accurate and cost-effective prediction model.

To get better results with less data and less processing, the ID-3 technique can be used. When dealing with enormous and continuous data volumes, the performance metric suffers. However, this is a computationally intensive process.

[37] state in their research that highly linked neurons are responsible for the brain's information processing. This method may be used to identify patterns and categorise data. In a computer network, each node is referred to as a "neuron." Connection nodes may be used to change neuron-to-neuron signals. Artificial neural networks may be fed nonlinear inputs as a source of information. Increasing or decreasing the weighted edge size has a direct effect on the intensity of the edge signals. When the nodes set a threshold value, only neurons with values greater than or equal to that value may send messages to one another. An artificial neuron may be represented as a multi-layered structure. Each layer's inputs are tampered with in some fashion. Signals frequently have to pass through multiple intermediary levels before reaching their ultimate destination. It is feasible to use artificial neural networks to mimic the human brain and solve issues. A growing number of applications for artificial neural networks are being developed in the fields of medical diagnostics, video and image recognition, and many more.

It is a binary linear non-probabilistic classification method: One or more target classes are created based on their features in a training model. The data pieces are represented as points in space. There appears to be a widening of the gap between the various kinds of commodities. New instances are assigned to the correct side of the chasm according to the target classes they have been assigned to. Non-linear classification may be performed by the support vector machine if the input datasets are unlabeled. Unsupervised learning is used by the support vector machine in order to cluster the data because there are no goal classes. Based on the functionality, new instances are added to existing clusters. Non-linear support vector machines (SVMs) are used to construct an effective recommendation system [38]. It is done in this manner so that each new instance of the input data model may be further categorised. The hyperplane separates one part of the hyperplane from the other. Divided into two groups, the classes sit on each side of the partitions. To put it another way, SVM divides the world into distinct classes. The target classes may now be neatly divided into subgroups using a hyperplane. When selecting a hyperplane, it is important that it accurately identifies the target class. Depending on the scenario, the hyperplane is selected. If there are any target classes, then three hyperplanes are necessary. In this case, the phrase "distance margin" is employed. Changing the margin distance measurements depending on the situation yields better results. SVM classification techniques such as the maximum margin classifier and the soft margin classifier may be used using margin distance data. This helps it categorize things more quickly. Most efficient classifiers in a range of disciplines are support vector machines (SVM). It can also be used to diagnose and treat cardiovascular disease. With a modest margin of separation between the two objects, this method works effectively in high-dimensional situations. Aside from samples, it works well with a wide range of other dimensions as well. Another drawback of this approach is that it requires a lot of time and memory. Noise and ambiguity in the data might also have an impact on performance measures, which may be worsened. While SVM has several flaws, its appropriate and efficient classification characteristics mean that it has a wide range of practical applications.

12.5.2 Result Analysis

NSL KDD [39] was the data set utilized in the experiment. The NSL-KDD dataset contains 24 different forms of assaults, and the data is either normal or one of them. Probe, DoS, R2L, and U2R are all attacks that fit into one of these four categories. The result comparison of these three classifiers is shown below in Figure 12.5.

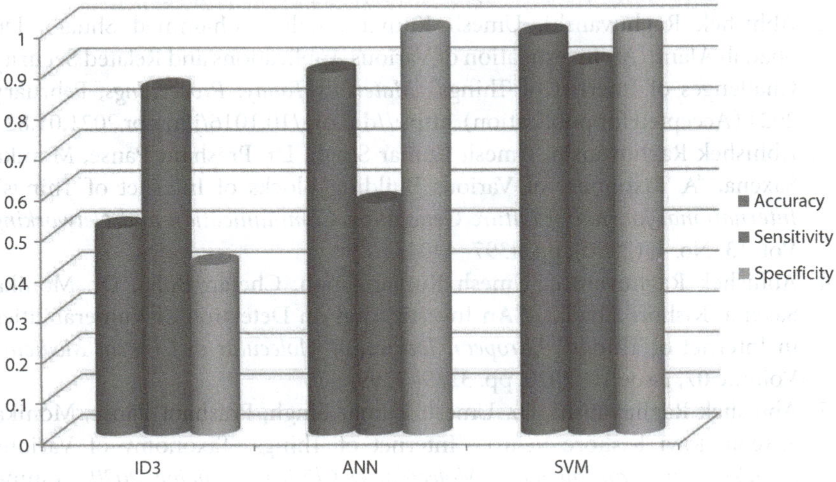

Figure 12.5 Result comparison of classifiers.

12.6 Conclusion

There is now a new type of IoT called IoV, which is when vehicles in a setting connect with many different things in society through the IoT. Fully autonomous cars and semi-autonomous cars exist together on the IoV. These vehicles may be able to reach both the outside and inside of their surroundings in the IoT. The cars communicate with other cars, the road, sensors, personal gadgets, pedestrians, and homes. Over the next few years, each car is likely to have more than 200 sensors in it. It will produce a lot of data. This data needs to be collected and processed correctly, so it must be done so. In the IoV, too, the safety of data is very important. This article is a detailed look at how artificial intelligence and machine learning can be used on IoV network data to predict all possible threats and protect IoV from a wide range of attacks and threats. This article talks about a machine learning-based intrusion detection system for the IoV. The KDD data set from NSL is used as the experimental data. There are three algorithms used in analytical research: the ANN, the SVM, and the ID3. SVM is better than ANN and ID3 when it comes to accuracy.

References

1. Abhishek Raghuvanshi, Umesh Kumar Singh, "Internet of Things for smart cities-security issues and challenges", *Materials Today: Proceedings*, December 2020 https://doi.org/10.1016/j.matpr.2020.10.849.
2. Abhishek Raghuvanshi, Umesh Kumar Singh, Mohammed Shuaib, Dr. Shadab Alam, "An Investigation of Various Applications and Related Security Challenges of Internet of Things", *Materials Today: Proceedings*, February 2021 (Accepted for publication) https://doi.org/10.1016/j.matpr.2021.01.821
3. Abhishek Raghuvanshi, Umesh Kumar Singh, Dr. Prashant Panse, Monika Saxena, "A Taxonomy of Various Building Blocks of Internet of Things", *International Journal of Future Generation Communication and Networking* Vol. 13, No. 4, (2020), pp. 4397–4404.
4. Abhishek Raghuvanshi, Umesh Kumar Singh, Chetan Bulla, Dr. Monika Saxena, Kishori Abadar, "An Investigation on Detection of Vulnerabilities in Internet of Things", *European Journal of Molecular & Clinical Medicine* Volume 07, Issue 10, 2020, pp. 3289–3299.
5. Abhishek Raghuvanshi, Dr. Umesh Kumar Singh, Prashant Panse, Monika Saxena, Ravi Kishore Veluri, "Internet of Things: Taxonomy of Various Attacks", *European Journal of Molecular & Clinical Medicine, 2020,* Volume 7, Issue 10, pp. 3853-3864.

6. Wu, Miao, *et al.* "Research on the architecture of Internet of Things." *Advanced Computer Theory and Engineering (ICACTE), 2010 3rd International Conference on*. Vol. 5. IEEE, 2010.
7. Saxena M. (2020). Novel framework and service model for internet of things in context of smart cities. *International Research Journal of Modernization in Engineering Technology & Science*. 2(9), 1730-1735.
8. Farooq MU, Waseem M, Khairi A, Mazhar S (2015), A critical analysis on the security concerns of Internet of Things (IoT). *Int J Comput Appl* 111:7.
9. Khan R, Khan S, Zaheer R, Khan S (2012), Future internet: the Internet of Things architecture, possible applications and key challenges. In: *2012 10th International Conference on Frontiers of Information Technology (FIT)*. IEEE, pp. 257–260.
10. Welch D, Lathrop S (2003), Wireless security threat taxonomy. In: *2003 IEEE Systems, Man and Cybernetics Society and Information Assurance Workshop*. IEEE, pp 76–83.
11. Ding Z-h, Li J-t, Feng B (2008), A taxonomy model of RFID security threats. In: *2008 11th IEEE International Conference on Communication Technology*. ICCT 2008. IEEE, pp. 765–768.
12. Yampolskiy M, Horvath P, Koutsoukos XD, Xue Y, Sztipanovits J (2013), Taxonomy for description of cross-domain attacks on CPS. In: *Proceedings of the 2nd ACM International Conference on High Confidence Networked Systems*. ACM, pp 135–142.
13. Mitrokotsa A, Rieback MR, Tanenbaum AS (2010), Classification of RFID attacks. Gen 15693:14443.
14. Khoo B (2011), RFID as an enabler of the Internet of Things: issues of security and privacy. In: *2011 International Conference on Internet of Things (ithings/CPSCom) and 4th International Conference on Cyber, Physical and Social Computing*. IEEE, pp 709–712.
15. Sastry AS, Sulthana S, Vagdevi S (2013), Security threats in wireless sensor networks in each layer. *Int J Advan Netw Appl* 4(4):1657.
16. Zhang W, Qu B (2013), Security architecture of the Internet of Things oriented to perceptual layer. *Int J Comput, Consum Control* (IJ3C) 2(2):37–45.
17. Douceur JR (2002), The Sybil attack. In: *International Workshop on Peer-to-Peer Systems*. Springer, pp. 251–260.
18. Ahmed N, Kanhere SS, Jha S (2005) The holes problem in wireless sensor networks: a survey. *ACM SIGMOBILE Mobile Comput Commun Rev* 9(2):4–18.
19. Padmavathi DG, Shanmugapriya M *et al.* (2009), A survey of attacks, security mechanisms and challenges in wireless sensor networks. arXiv:0909.0576
20. Cho J-S, Yeo S-S, Kim SK (2011), Securing against brute-force attack: a hash-based RFID mutual authentication protocol using a secret value. *Comput Commun* 34(3):391–397.
21. Mattern F, Floerkemeier C (2010), From the internet of computers to the Internet of Things, From active data management to event based systems and more, in *Lecture Notes in Computer Science*, pp 242–259.

22. Zhu B, Joseph A, Sastry S (2011), A taxonomy of cyber attacks on SCADA systems. In: *Internet of Things (Ithings/CPSCom), 2011 international conference on and 4th international conference on Cyber, Physical and Social Computing*. IEEE, pp 380–388
23. Jia YJ, Chen QA, Wang S, Rahmati A, Fernandes E, Mao ZM, Prakash A, Unviersity SJ (2017), ContexIoT: towards providing contextual integrity to appified IoT platforms. In: *Proceedings of the 21st Network and Distributed System Security Symposium (NDSS'17)*.
24. Fernandes E, Jung J, Prakash A (2016), Security analysis of emerging smart home applications. In: *2016 IEEE Symposium on Security and Privacy (SP)*. IEEE, pp 636–654.
25. Thakur BS, Chaudhary S (2013) Content sniffing attack detection in client and server side: a survey. *Int J Advan Comput Res* 3(2):7.
26. Simmons C, Ellis C, Shiva S, Dasgupta D, Wu Q (2009), AVOIDIT: a cyber attack taxonomy.
27. Bugiel S, Heuser S, Sadeghi A-R (2013), Flexible and fine-grained mandatory access control on Android for diverse security and privacy policies. In: *USENIX Security Symposium*, pp. 131–146.
28. Jensen M, Schwenk J, Gruschka N, Iacono LL (2009), On technical security issues in cloud computing. In: *2009 IEEE International Conference on Cloud Computing. CLOUD'09*. IEEE, pp. 109–116.
29. Zhang Q, Wang X (2009) SQL injections through back-end of RFID system. In: *2009 International Symposium on Computer Network and Multimedia Technology. CNMT 2009*. IEEE, pp. 1–4.
30. Garca, S.; Grill, M.; Stiborek, J.; Zunino, A., An empirical comparison of botnet detection methods. *Computer Security*. 2014, 45, 100–123.
31. Divekar, A.; Parekh, M.; Savla, V.; Mishra, R.; Shirole, M.Benchmarking datasets for Anomaly-based Network IntrusionDetection: KDD CUP 99 alternatives. *IEEE 3rd International Conference on Computing, Communication and Security (ICCCS)*, Kathmandu, Nepal, 25–27 October 2018; pp. 1–8.
32. Belouch, M.; El Hadaj, S.; Idhammad, M., Performanceevaluation of intrusion detection based on machine learning using Apache Spark. *Procedia Comput. Sci.* 2018, 127, 1–6.
33. S. Abdufattokhov and B. Muhiddinov, "Probabilistic Approach for System Identification using Machine Learning," in *2019 International Conference on Information Science and Communications Technologies (ICISCT)*, 2019, pp. 1–4, doi: 10.1109/ICISCT47635.2019.9012025.
34. S. Abdufattokhov and B. Muhiddinov, "Stochastic Approach for System Identification using Machine Learning," in *2019 Dynamics of Systems, Mechanisms and Machines (Dynamics)*, 2019, pp. 1–4, doi: 10.1109/Dynamics47113.2019.8944452.
35. Priyanka Sharma, Dr. N. Vasantha Gowri, Dr. Mohd Naved, Bharat M. N. (2021). An Internet of Things for prevention of security attack on cloud

medical data using artificial intelligence. *International Journal of Future Generation Communication and Networking*, 14(1).

36. A. Rajeshkanna and K. Arunesh, "ID3 Decision Tree Classification: An Algorithmic Perspective based on Error rate," in *2020 International Conference on Electronics and Sustainable Communication Systems (ICESC)*, 2020, pp. 787-790, doi: 10.1109/ICESC48915.2020.9155578.
37. T. H. F. Harumy, M. Zarlis, S. Effendi and M. S. Lidya, "Prediction Using A Neural Network Algorithm Approach (A Review)," in *2021 International Conference on Software Engineering & Computer Systems and 4th International Conference on Computational Science and Information Management (ICSECS-ICOCSIM)*, 2021, pp. 325-330, doi: 10.1109/ICSECS52883.2021.00066.
38. J. Huang, J. Zhou and L. Zheng, "Support Vector Machine Classification Algorithm Based on Relief-F Feature Weighting," in *2020 International Conference on Computer Engineering and Application (ICCEA)*, 2020, pp. 547-553, doi: 10.1109/ICCEA50009.2020.00121.
39. http://nsl.cs.unb.ca.

13

ML Techniques for Attack and Anomaly Detection in Internet of Things Networks

Vinod Mahor[1]*, Sadhna Bijrothiya[2], Rina Mishra[3] and Romil Rawat[3]

[1]Department of Computer Science & Engineering, IES College of Technology, Bhopal, M.P., India

[2]Department of Computer Science & Engineering, Maulana Azad National Institute of Technology, Bhopal, MP, India

[3]Department of Computer Science & Engineering, Shri Vaishnav Vidyapeeth Vishwavidyalaya, Indore, India

Abstract

The Internet of Things (IoT), is a fundamental driver of smart cities. It is the champion of the global collaboration of machines/things, people, huge data, and processes to create cities that are efficient, economically feasible, and human-friendly. It is a technique that connects thousands of autonomous devices to accumulate different types of data and their circumstances via external devices. With this they can share the information with authorized personnel to serve various purposes including organizing and coordinating economic services and developing commercial services and activities. On the other side, the Internet of Things is currently experiencing more security challenges than ever before. Machine Learning (ML) has made major technological progress, offering up a slew of new scientific avenues to solve existing and future IoT concerns. It is an effective technique for identifying dangers and suspicious acts carried out by sophisticated gadgets and networking in clever wired and wireless networks. This article is mainly focused on ML approaches and the relevance of IoT security in the framework of different types of probable assaults. Alternative ML-based IoT security approaches have also been suggested.

Keywords: Anomaly-detection, cyber-attacks, Internet of Things (IoT), Machine Learning (ML)

**Corresponding author*: vinodengg.mt@gmail.com

Romil Rawat, A. Mary Sowjanya, Syed Imran Patel, Varshali Jaiswal, Imran Khan and Allam Balaram (eds.) Autonomous Vehicles Volume 1: Using Machine Intelligence, (235–252) © 2023 Scrivener Publishing LLC

13.1 Introduction

IoT is a technique to connect electrical devices to a server and exchange data with it without human intervention [1–3]. But the security of the IoT system is a major problem. The increasing number of intelligent gadgets containing private and confidential information nowadays is the major issue of concern [2].

As with most IoT devices, devices have limited resources (e.g., batteries). Bandwidth, storage, and computation; extremely customizable and sophisticated Algorithm-based protection solutions are ineffective. ML and DL algorithms have been widely employed for a variety of tasks such as IoT applications, regression, and classification such as intrusion detection, image analysis, and recommendation systems [4, 5]. Machine learning (ML) techniques are a fantastic solution for ensuring the stability of IoT solutions. ML is a revolutionary artificial intelligence technology that is not so difficult and can outperform complex networks as well [6, 7]. ML approaches are used to train a machine and create a variety of assaults and a safety plan. Furthermore, advances in Machine Learning show promise in recognizing and intelligently addressing emerging dangers via learning abilities. Future IoT device security protocols would also improve the reliability and accessibility of ML algorithms.

This article is organized as follows: Section 13.2 gives a summary of IoT, particularly security issues, levels, and the relevance of IoT. Section 13.3 discusses IoT assaults, their effect and multiple attack surfaces, detection and classification in IoT, machine learning in security measures, and other sorts of classification methods, and IoT security solutions [7, 8]. Section 13.4 offers an overview of published studies on ML security, and Section 13.5 gives a conclusion.

13.2 Internet of Things

The Internet of Things (IoT), refers to billions of tangible items connected to the Web, as well as the global storage and data exchange. With the advent of low-cost computer chips and a widespread wireless network, everything from a pin to an airplane may now be converted into an IoT component. By connecting and adding sensors to all of these diverse items, artificial intelligence is applied to gadgets, allowing them to communicate real-time data without requiring the assistance of a person. The Internet of Things improves our society's intelligence and adaptability by fusing the digital

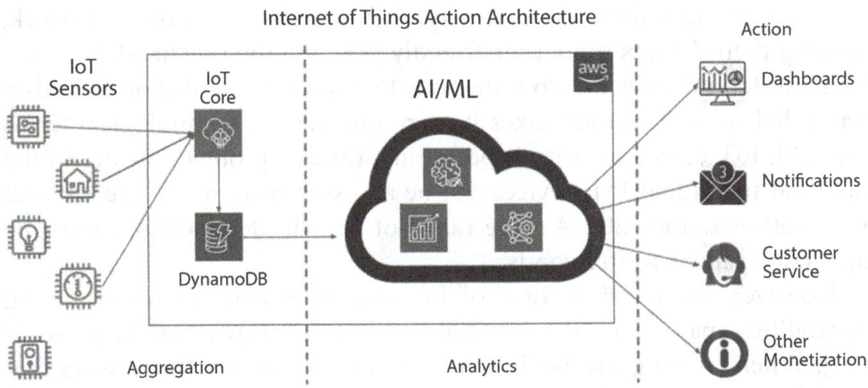

Figure 13.1 IoT in action.

and physical worlds. The Figure 13.1 represents modelling of IoT framework in action.

A. Layers of IoT
The IoT architecture serves as a portal to multiple hardware/software application-programmers to create a link and enhance IoT technologies at each part of entry [9]. To send and receive information/data from different layers of the IoT architecture, several networking protocols such as Bluetooth, Wi-Fi, RFID, narrow and broadband, ZigBee, and LPWAN are used [10, 11]. A basic IoT architecture is made up of three layers: physical, network, and application [12, 13]. We have been particularly ambiguous regarding the kind of data created by IoT devices, as well as the nature of data processing. Cloud computers process data in a big centralized form in various system topologies. A cloud-centric design places the cloud in the center, with apps above it and a network of smart things underneath it [14]. Cloud computing is prioritized because it offers exceptional springiness and scaling. It offers storage, services, infrastructural facilities, a framework, and software. Cloud computing may be used by builders to supply storage technology and software services, "data mining" and "machine learning" tools, and "visualization" techniques.

B. Security in IoT
The security of IoT devices has been a hot topic in the 21st century; on the one hand, IoT connects and brings the entire universe closer together. On the other hand, it offers up new avenues for various types of assaults [15, 16].

IoT apps are utilized for a variety of functions via an open network, making their devices more user-friendly [17]. On the one hand, IoT puts human life in danger due to numerous threats and assaults; on the other hand, IoT simplifies and makes it more obedient in technological terms [18, 19]. IoT device security is becoming a major problem, owing to the fact that individual IoT devices can be accessed from anywhere without user authorization [20]. A wide range of security technologies must be applied to safeguard IoT goods.

However, the physical form of IoT devices restrict their computing capabilities, making it difficult to build a comprehensive security protocol [21]. When creating a stable IoT, it is critical to examine the features that constitute protection. Security standards in a conventional IoT application are divided into three primary categories: (i) secrecy, (ii) integrity, and (iii) authentication [22]. Figure 13.2 represents the architecture of IoT layers.

- Confidentiality means keeping information hidden from third parties. Sensitive sensors, for example, require confidentiality when dealing with sensitive military material, which is why one of the most often requested features is the Wireless Sensor Network (WSN) technology [23].
- To maintain the integrity of IoT data, the communication receiver must confirm that messages received during transmission or delivery have not been altered. In other words, the integrity of the data guarantees that the provided data has not been altered or corrupted. This is especially important because even if hackers cannot get data, the network may not function properly, if compromised nodes corrupt the supplied data. Indeed, if the communication link is unreliable, data can be updated without the presence of an intruder. Integrity control ensures that both accidental and intentional alterations to the message are detected [25].

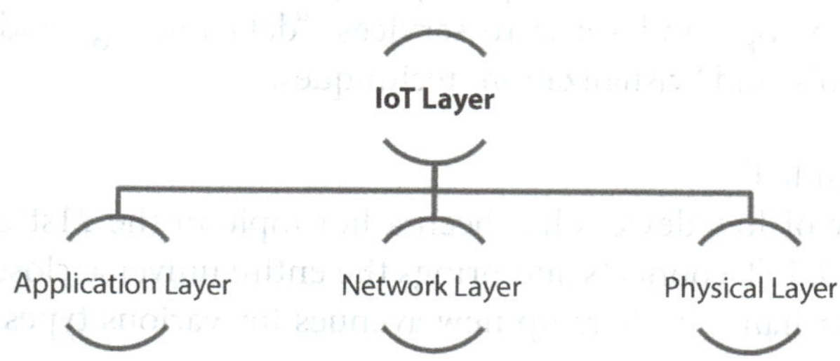

Figure 13.2 The architecture of IoT layers.

- The authentication process verifies from where the communication originated, where it is claimed or is what it is purported to be. The sensor nodes must establish the identity and validity of the peer node they are communicating with. Authenticity guarantees that the message is genuine. Message Authentication Code (MAC) is a short piece of information used for message authentication that ensures the integrity and authenticity of the message [26, 27].

13.3 Cyber-Attack in IoT

Various assaults on the IoT system have occurred in recent years, increasing manufacturers' and users' awareness of IoT goods. This section describes the many kinds of assaults, consequences, and IoT surfaces.

There are two types of IoT attacks: cyber and physical. As shown in Figure 13.3, cyber-attacks encompass both passive and aggressive attacks. A cyber-attack threat targets various IoT devices by hacking and wireless network activities (stealing, wiping, modifying, or destroying) the user's data. Actual attacks, on the other hand, lead to IoT devices suffering physical damage.

There is no need for a network in this case to attack the device. Physical IoT devices, such as mobile devices, cameras, sensors, routers, and so on, are also vulnerable to such attacks [28, 29].

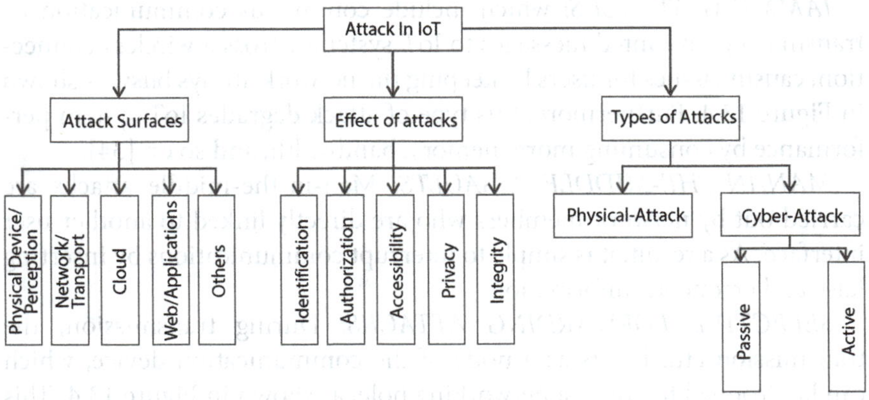

Figure 13.3 Complete list of IoT attacks, including various attacks, attack surfaces and attack effects.

A. Active Attacks

When a network attacker accesses the interface settings and disconnects particular services, this is considered an active assault [28]. IoT device security may be compromised in a variety of methods, including interruption, intervention, and changes in ongoing assaults, jamming, active attacks, and hole attacks inputs that are malicious, assaults from the middle, forwarding with discretion, spoofing, as in DoS, Sybil launches an assault, data manipulation, and so on, are depicted in Figure 13.4 [29].

Denial of Service (DoS) Attacks: As shown in Figure 13.4, DoS attacks are largely responsible for disabling system services by producing a large number of repeating demands. As a result, the user is unable to access and connect to the IoT device, preventing informed decision-making.

Furthermore, DoS attacks keep IoT devices always powered on, limiting battery life [30]. A distributed denial of service (DDoS) assault happens when many attacks are launched from different IP addresses to produce a variety of requests in order to keep the server busy. This makes differentiating between legitimate and malicious traffic impossible [31]. Along with current history, a specific "IoT" .net virus known as Mirai has been responsible for launching disruptive DDoS-attack wreaking havoc on thousands of IoT system computers, making them malfunction as a result of interference [32].

Spoofing and Sybil attacks: as shown in Figure 13.4, these types of attacks are mostly employed to get unauthorized access to IoT devices by targeting user identity (RFID and MAC address). IoT systems are especially prone to spoofing attacks because TCP/IP suite lack a complete security mechanism.

Furthermore, there are two attacks launched such as man-in-the-middle attacks and denial-of-service (DoS) attacks [33].

JAMMING ATTACKS: which include continuous communication by transmitting unwanted messages to IoT systems across a wireless connection, causing issues for users by keeping the network always busy, as shown in Figure 13.4. Furthermore, this type of attack degrades IoT system performance by consuming more memory, bandwidth, and so on [34].

MAN-IN-THE-MIDDLE ASSAULTS: Man-in-the-middle attacks are carried out by network members who are directly linked to another user interface. As a result, it is simple to interrupt communications by inserting false and erroneous information.

SELECTIVE FORWARDING ATTACKS: During transmission, the transmission attack acts as a node of the communication device, which can be dropped to create a networking hole, as shown in Figure 13.4. This type of assault is difficult to identify and stop [35].

BLACK-HOLE AND GRAY-HOLE: attacks are classed as Active assaults since they influence network performance and lead the network to collapse, as shown in Figure 13.4 [36].

DATA TAMPERING: Data tampering not only endangers companies but also endangers people's lives and property. As a result, businesses must take efforts to avoid such attacks and mitigate any harm they may do, as seen in Figure 13.4 [36].

B. Passive Attack in IoT

Passive attacks are intended to gather information about the user without their awareness and decrypt their private, encrypted data. The two most common ways for performing a passive attack on an IoT network are eavesdropping and traffic monitoring [37].

Snooping: The attacker listens in on messages sent between two entities. For the attack to be effective, the traffic must not be encrypted. Any unencrypted information, such as a password submitted in response to an HTTP request, can be obtained by the attacker [36, 37].

Traffic analysis: The attacker studies the metadata transmitted in traffic to derive traffic information, such as the exchanged traffic and the entities involved (rate, duration, and so on). If encrypted data is used, traffic analysis can also lead to cryptanalysis attacks, resulting in the attacker acquiring information or successful unencrypted traffic [38].

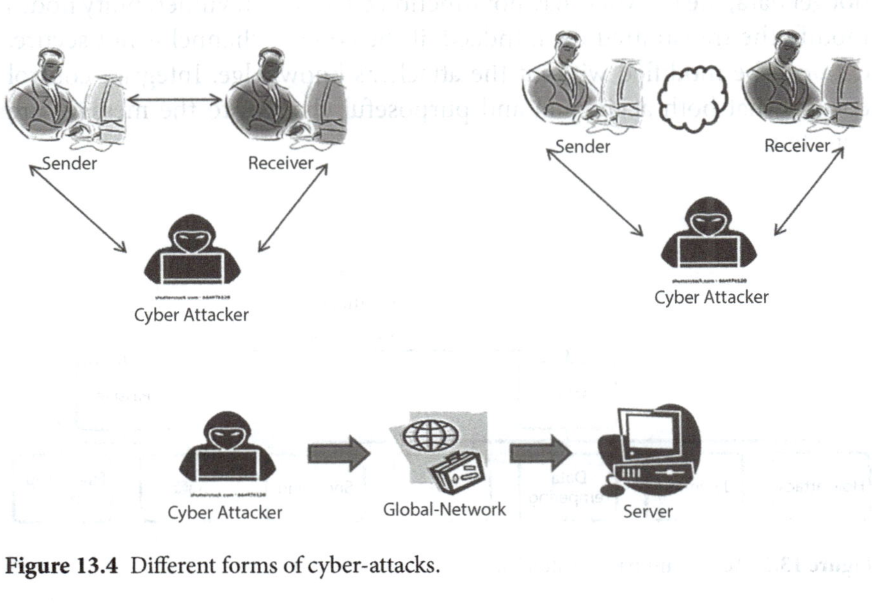

Figure 13.4 Different forms of cyber-attacks.

C. The Effects of Cyber-Attacks on IoT

IoT threats pose a danger to the network's ability to secure users' confidentiality, validation, and permission. Figure 13.5 depicts a comprehensive list of numerous kinds of attacks, as well as their consequences on IoT devices. The following elements must be addressed while creating a security protocol for IoT device attacks [37].

Accessibility ensures that IoT device services are always delivered to authorized users. At the same time as the creation of an actual "IoT" network is critical, jamming attacks and DoS weaken our services simply developing unjustified demands and keeping its network is overloaded. The compact security protocols are required to make IoT systems available to user clients without interruption [38, 39].

Privacy is the sole aspect of the IoT architecture that contains both active and passive assaults. Currently, individual data, medicinal data, and nationwide security information, confidential data, is securely encrypted and delivered over the Internet via different IoT devices that are not intended to be accessed by illegal clients [39].

Since hackers may track down and decrypt the physical address of IoT devices, the majority of the information is easy to conceal from unauthorized third parties [40].

INTEGRITY: To safeguard data integrity in the IoT, the receiver must ensure that the messages received during transmission have not been modified. The data integrity verifies that the data supplied has not been updated or altered [39, 40]. It is extremely important because, even if hackers cannot get data, the network may not function effectively if vulnerability nodes modify the transmitted data. Indeed, if the contact channel is not secure, data can be modified without the attacker's knowledge. Integrity control ensures that both accidental and purposeful changes to the message are detected.

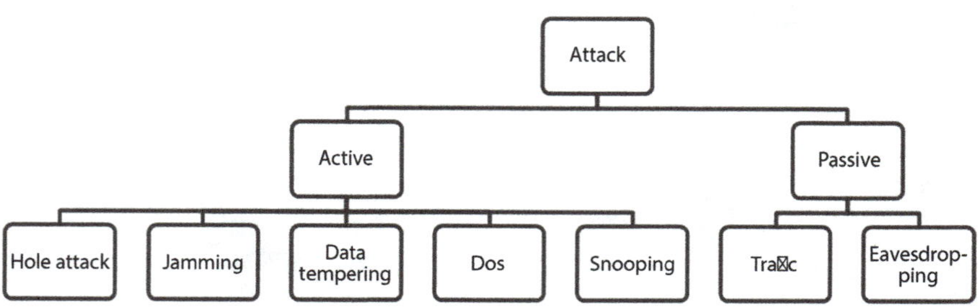

Figure 13.5 Active and passive attacks.

IDENTIFICATION: Identification refers to the user's IoT network authorization. Clients must initially be registered in order to connect with the Cloud Server, although the commercialization and robustness of IoT systems pose identification challenges. Sybil and spoofing attacks compromise network security, allowing access to attackers without valid identity, and they gain access on server. To check this problem, an efficient IoT system is required.

AUTHORIZATION: This refers to the user's access to the IoT system. It only allows authorized clients to enter, track, and use data from IoT networks. The commands of system users with authority are also performed. It is difficult to preserve every customer record and grant anyone permission depending on the data since people are human, but sensors, equipment, and services are not.

Identification refers to the user's permission in the IoT network. Clients must initially register in order to interact with both the cloud platform. IoT technology trade-offs with resilience, on the other hand, make detection difficult, and spoofing attacks are to blame for weakening network security, with attackers gaining rapid login to the server in the absence of sufficient identity. As a result, an actual IoT system identify strategy is required to offer critical security while adhering to system restrictions.

D. Detecting Anomalies in IoT Attacks

There are instances of real-world data sets that are unique to all others and are identified as anomalies. Identifying anomalies is the process of locating phenomena that are considered abnormal in their activity in comparison to normal nodes. The intrusion protection system, fraud detection, and data leaking are all separate causes of abnormalities. Anomaly detection is applied in a variety of IoT sectors, including smart cities, network security, industries, and so on [13–15]. The Figure 13.6 represents classification of different anomaly frameworks.

INTRUSION DETECTION: IoT devices are Internet-interconnected devices which are vulnerable to security assaults. DoS and DDOS attacks do substantial damage to the IoT network. The most difficult problem in IoT deployments is identifying and preventing such threats [10, 11].

FRAUD DETECTION: During logins or online payments, IoT nets persist vulnerable to intercepting info about credit and debit cards, bank information and other individual information [10–12].

DATA LEAKAGE: External firms might leak sensitive data from databases, file servers, and other information sources, resulting in not just information loss but also a danger to secrecy. Such leaks can be avoided with proper encryption measures [23].

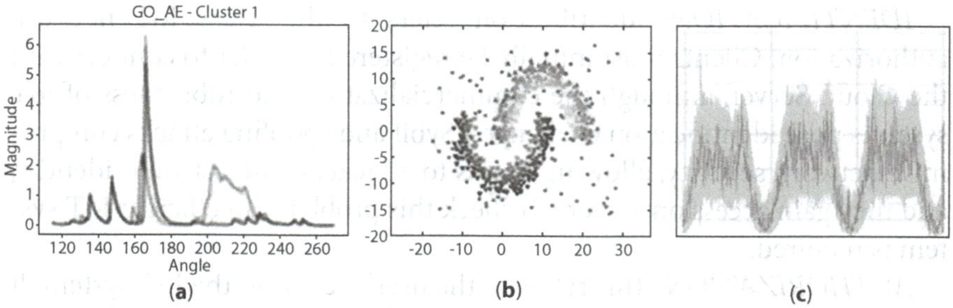

Figure 13.6 Clarification of different type point wise, cooperative, prescribed anomaly.

13.4 IoT Attack Detection in ML Technics

ML is a type of machine learning and a field of AI that trains machines utilizing different tricks and allows robots to learn through their interactions rather than being unambiguously programmed [36]. ML does not require human aid, sophisticated math techniques, or performance in complex networks.

In recent years, ML methods for IoT security have improved significantly. ML techniques may thus foresee diverse IoT scenarios based on an early assessment of system behavior. Furthermore, by integrating more machine learning algorithms, viable solutions to resource-limited IoT devices may be developed. Every article is broken into two sections, parts machine learning method and machine learning with IoT security technologies [36, 37, 40].

TECHNIQUES FOR MACHINE LEARNING (ML): ML approaches with "supervised learning", "unsupervised learning", and "reinforcement learning" can identify novel attacks in IoT devices and begin a solid protection plan. Figure 13.7 demonstrates several machine learning methods used to secure IoT devices.

Supervised learning is the most often utilized learning approach in machine learning, employing a qualification process, set of data and a statistical model, the outcome is marked input and produce. "Classification" or "regression" training are two types of supervised learning [24]. In unsupervised learning, there seems to be no o/p data to those input data. The majority of the data is unmarked, and the algorithm attempts to discover correlations between this data collection. It categorises them as clusters of different kinds [38].

Furthermore, Reinforcement learning enables the Natural language processing (NLP) interactions and relationships with its atmosphere in

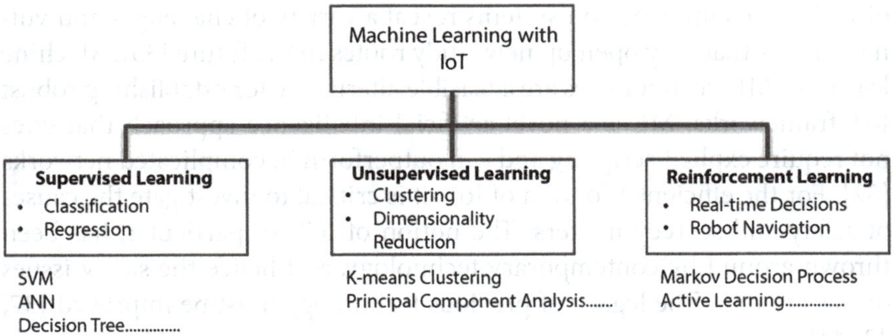

Figure 13.7 Classify the category of machine learning.

the same manner that people do by doing actions that optimize overall reaction. The reaction might be a price dependent on the outcome of the operation. In reinforcement learning, there are no predetermined behaviors for every specific operation, and the system employs hearing then fault approaches. By test and fault, the agent may discover and use the optimum strategy from its knowledge to get the maximum remuneration.

MACHINE LEARNING-BASED IoT ATTACK DETECTION SOLUTION: In recent years, the topic of ML and Internet of Things vulnerability scanning has emerged as a study area. This section has examined several ML methods as potential possibilities for safeguarding IoT devices. These methods were investigated utilizing an IoT structure's three key architectural layers, the physical/perception layer, the network layer, and the web/application layer [31].

Traditional authentication systems for securing the physical surface are inadequate owing to the precise threshold value required to detect undesired signals that cause false alarms. As a result, ML-based learning techniques might be employed as a physical layer authentication alternative [40]. QL-based learning approach reduces validation fault by around 64.3 percent and displayed superior enactment than typical authentication m layers.

Other studies have been conducted to set parameters for the logistical regression model for supervised learning approaches such as Francis Wolfe Distribution with Sequential Consolidated Hill in order to reduce overhead communication and improve spoofing detection performance [41].

IoT NETWORK-SECURITY AND ML-TECHNIQUES' CHALLENGES AND GAPS: Nowadays, the Internet of Things (IoT) and its significance have entered every home. Furthermore, the security of IoT has caught the interest of a large number of network and device scholars. The deployment of IoT and its influence on systems reveal a variety of challenges and vulnerabilities that may open up new study routes in the future [33]. Machine learning (ML) technologies are a feasible alternative for establishing robust IoT frameworks. ML is a novel artificial intelligence approach that does not require explicit scripting and can outperform in complicated networks [32]. For the efficient adoption of IoT, it is critical to investigate the causes of safety and secrecy matters. The notion of IoT, in particular, has been thrown around by contemporary technology, and hence the safety issues of IoT are new. The legacy of previous technology must be improved [37, 42–44].

Focused on the similarities and differences between IoT and security issues, they have concentrated on privacy-related issues. Software, electronics, networks, and applications are the primary drivers of similarities and differences. According to these categories, security issues in the traditional IT domain are virtually comparable to those in the IoT domain. However, the fundamental issue of the IoT is resource limits, which impede the use of advanced security solutions that are now accessible on IoT networks. Furthermore, IoT security and privacy technologies necessitate cross-cutting layer architecture and optimized algorithms. Due to computational constraints, IoT systems may require the development of new breeds of optimized cryptology and other procedures for safety and secrecy.

EVALUATION AND RECOMMENDATIONS: As shown in Table 13.1, researchers applied several ML algorithms and strategies to identify Attacks and Anomalies, and they achieved excellent results in accuracy detection. Researchers used and compared numerous ML algorithms in studies [19, 27–31], and the Random Forest (RF) algorithm produced the best results with 99.34 percent, 99.5 percent, 99.4 percent, 99.9 percent, 99 percent, 99.5 percent, and 99.9 percent accuracy. While studying [36]. The findings reveal that DT and KNN performed better than the other algorithms; nevertheless, when compared to the DT method, the KNN takes much longer to classify.

Furthermore, in experiments [20], the combination of the two ML algorithms, RF and DT, resulted in improved accuracy for identifying Attacks. According to research [33], the two ML algorithms, RF and KNN, achieved 99 percent accuracy in identifying assaults. Overall, based on the reviewed research, it was revealed that the Random Forest ML method performs well in identifying Attacks and Anomalies.

ML has proven its worth in general cyber-security applications and is suited for dealing with a wide range of IoT-related concerns. They balance a wide range of IoT network risks depending on the speed with which ML-based systems response and their adaptability. ML research for many

Table 13.1 Comparing the performance of Ml techniques for intrusion detection system and cyber-attack detection in IoT environments.

Year	Author	Objectives	Datasets	Results and accuracy	Techniques
2019	Hasan *et al.* [19]	The efficiency of many machine learning models has been carefully compared to predict attacks and anomalies on IoT networks.	NSL-KDD, Real Traffic, DS2OS	Random Forest has recorded the best accuracy 99.4%	LR, DT, RF, SVM, and ANN
2020	Mary DRTa SAS [29]	serviceability issues with ML for detecting anomalies in IoT networks.	IoT-23	RF algorithm achieved the best results with an accuracy of 99.5%	SVM, RF, Naïve Bayes, Decision Tree
2020	Liu *et al.* [28]	Enhance IoT security through the experimentation of multiple machine learning methods on the IoT network intrusion dataset.	IoT Network Intrusion Dataset	the accuracy when using KNN algorithms was 99%	LR, SVM, RF, KNN, XGBoost

(Continued)

Table 13.1 Comparing the performance of Ml techniques for intrusion detection system and cyber-attack detection in IoT environments. (*Continued*)

Year	Author	Objectives	Datasets	Results and accuracy	Techniques
2020	Al-Akhras *et al.* [27]	examines various ML algorithms' effectiveness to detect Attacks and anomaly in IoT Networks	UNSW-NB15	RF and KNN classifiers perform best with 100% accuracy without noise injection and 99% accuracy with 10% noise filtering	RF, KNN, Naïve bayes
2020	Rani and Kaushal [30]	Improve the security and accuracy of Intrusion Detection System (IDS)	NSL-KDD and KDDCUP99	The proposed simulation has a 99.9% accuracy with less time and energy intrusion detection	KNN, NB, Decision Tree, Logistics Regression, RF
2020	Stoian [31]	Anomaly Detections and Attacks in IoT Networks.	IoT-23	The RF algorithm has obtained the best results with 99.5% accuracy	RF, NB, MLP, SVM, and AdaBoost
2021	Bagui *et al.* [36]	Intrusion and attack detection for IoT Botnet.	UCI's machine learning repository	Using these three ML classifiers was over 99% in most cases and 100% in many cases.	LR, SVM, RF

sorts of applications is thriving. There are several data points that demonstrate the importance of ML as an emerging technology.

13.5 Conclusion

By bringing global topics in our hands, the IoT has the ability to revolutionize the world. As a consequence of the blessings of IoT-based smart services, anyone may acquire, link, and store information anywhere across the network. IoT improves our lives by linking us to the virtual globe to make work simpler, faster, and more efficient. This chapter conducts a "systematic review of the literature" on "Machine Learning-based IoT security", including a study of IoT and its architecture, a detailed examination of several types of security threats, many machine learning algorithm based more categories, and machine learning algorithm based security solutions. Additionally, Machine Learning algorithms must have been researched in terms of potential problems, which might assist future researchers in determining their ultimate goals and achieving them.

References

1. Fawzi LM, S. Y. Ameen, S. A. Dawwd, and S. M. Alqaraawi, "Comparative Study of Ad-hoc Routing Protocol for Oil and Gas Pipelines Surveillance Systems," *International Journal of Computing and Network Technology*. 2016;4.
2. Xiao L, Y. Li, G. Han, G. Liu, W. Zhuang, "PHY-layer spoofing detection with reinforcement learning in wireless networks," *IEEE Transactions on Vehicular Technology*. 2016;6510037–10047.
3. Saied A, Overill RE, Radzik T. "Detection of known and unknown DDoS attacks using Artificial Neural Networks," *Neurocomputing*. 2016;172385–393.
4. Oracevic A, S. Dilek, S. Ozdemir, "Security in internet of things A survey," in *2017 International Symposium on Networks, Computers and Communications (ISNCC)*. 2017;1–6.
5. Yang Y, L. Wu, G. Yin, L. Li, and H. Zhao, "A survey on security and privacy issues in Internet-of-Things," *IEEE Internet of Things Journal*. 2017; 41250–1258.
6. Xiao L, X. Wan, Z. Han, "PHY-layer authentication with multiple landmarks with reduced overhead," *IEEE Transactions on Wireless Communications*. 2017;171676–1687.
7. Adat V, Gupta B. "Security in Internet of Things Issues, challenges, taxonomy and architecture," *Telecommunication Systems*. 2018;67423–441.

8. Alam MS, D. Husain, S. Naqvi, P. Kumar, "IOT security through Machine Learning and homographic encryption technique," in *International Conference on New Trends in Engineering & Technology (ICNTET)*, Chennai; 2018.
9. Ali ZA, Ameen SY. "Detection and Prevention Cyber-Attacks for Smart Buildings via Private Cloud Environment," *International Journal of Computing and Network Technology*. 2018;627–33.
10. Ameen SY, Ali ALSH. "A Comparative Study for New Aspects to Quantum Key Distribution," *Journal of Engineering and Sustainable Development*. 2018;1145–57.
11. Baracaldo N, B. Chen, H. Ludwig, A. Safavi, and R. Zhang, "Detecting poisoning attacks on machine learning in iot environments," in *2018 IEEE international congress on internet of things (ICIOT)*. 2018;57–64.
12. Benkhelifa E, Welsh T, Hamouda W. "A critical review of practices and challenges in intrusion detection systems for IoT Toward universal and resilient systems," *IEEE Communications Surveys & Tutorials*. 2018;203496–3509.
13. Janaby AO. al, A. Al-Omary, S. Y. Ameen, and H. M. Al-Rizzo, "Tracking High-Speed Users Using SNR-CQI Mapping in LTE-A Networks," in *2018 International Conference on Innovation and Intelligence for Informatics, Computing, and Technologies (3ICT)*. 2018;1–7.
14. Alrashdi I, Alqazzaz A, Aloufi E, Alharthi R, Zohdy M, Ming H. "Ad-iot Anomaly detection of iot cyberattacks in smart city using machine learning," in *2019 IEEE 9th Annual Computing and Communication Workshop and Conference (CCWC)*. 2019;0305–0310.
15. Alsamiri J, Alsubhi K. "Internet of Things cyber attacks detection using machine learning," *Int. J. Adv. Comput. Sci. Appl.* 2019;10.
16. Al-Sultan MR, Ameen SY, Abduallah WM. "Real Time Implementation of Stegofirewall System," *International Journal of Computing and Digital Systems*. 2019;8498–504.
17. Arko AR, Khan SH, A. Preety, M. H. Biswas, "Anomaly detection In IoT using machine learning algorithms," Brac University; 2019.
18. Costa KA. da, J. P. Papa, C. O. Lisboa, R. Munoz, and V. H. C. de Albuquerque, "Internet of Things A survey on machine learning-based intrusion detection approaches," *Computer Networks*. 2019;151147–157.
19. Hasan M, Islam MM, Zarif MII, Hashem M. "Attack and anomaly detection in IoT sensors in IoT sites using machine learning approaches," *Internet of Things*. 2019;7100059.
20. Elazhary H. "Internet of Things (IoT), mobile cloud, cloudlet, mobile IoT, IoT cloud, fog, mobile edge, and edge emerging computing paradigms Disambiguation and research directions," *Journal of Network and Computer Applications*. 2019;128105–140.
21. Sharma B, Sharma L, Lal C. "Anomaly Detection Techniques using Deep Learning in IoT A Survey," in *2019 International Conference on Computational Intelligence and Knowledge Economy (ICCIKE)*. 2019;146–149.

22. Singh A, A. Payal, S. Bharti, "A walkthrough of the emerging IoT paradigm Visualizing inside functionalities, key features, and open issues," *Journal of Network and Computer Applications*. 2019;143111–151.
23. Hassija, V. Chamola, V. Saxena, D. Jain, P. Goyal, and B. Sikdar, "A survey on IoT security application areas, security threats, and solution architectures," *IEEE Access*. 2019;782721–82743.
24. Zebari IM, Zeebaree SR, Yasin HM. "Real Time Video Streaming From Multi-Source using Client-Server for Video Distribution," in *2019 4th Scientific International Conference Najaf (SICN)*. 2019;109–114.
25. Aysa MH, Ibrahim AA, Mohammed AH. "IoT ddos attack detection using machine learning," in *2020 4th International Symposium on Multidisciplinary Studies and Innovative Technologies (ISMSIT)*. 2020;1–7.
26. Elmrabit N, Zhou F, F. Li, H. Zhou, "Evaluation of machine learning algorithms for anomaly detection," in *2020 International Conference on Cyber Security and Protection of Digital Services (Cyber Security)*. 2020;1–8.
27. Al-Akhras M, M. Alawairdhi, A. Alkoudari, and S. Atawneh, "Using machine learning to build a classification model for iot networks to detect attack signatures." 2020.
28. Liu Z, Thapa N, A. Shaver, K. Roy, X. Yuan, S. Khorsandroo. "Anomaly Detection on IoT Network Intrusion using Machine Learning," in *2020 International Conference on Artificial Intelligence, Big Data, Computing and Data Communication Systems (icABCD)*. 2020;1–5.
29. Mary DRTaSAS. "Attack and Anomaly Detection in IoT Networks using Machine Learning," *Int. J. Comput. Sci. Mob. Comput*. 2020;995–103.
30. Rani D, Kaushal NC. "Supervised Machine Learning Based Network Intrusion Detection System for Internet of Things," in *2020 11th International Conference on Computing, Communication and Networking Technologies (ICCCNT)*. 2020;1–7.
31. Stoian N-A. "Machine Learning for anomaly detection in IoT networks Malware analysis on the IoT-23 data set," University of Twente; 2020.
32. Liu Z, Thapa N, A. Shaver, K. Roy, X. Yuan, S. Khorsandroo. "Anomaly Detection on IoT Network Intrusion using Machine Learning," in *2020 International Conference on Artificial Intelligence, Big Data, Computing and Data Communication Systems (icABCD)*. 2020;1–5.
33. Abdulqadir HR, Zeebaree SR, Shukur HM, M. M. Sadeeq, B. W. Salim, A. A. Salih, *et al*., "A study of moving from cloud computing to fog computing," *Qubahan Academic Journal*. 2021;160–70.
34. Ageed ZS, Zeebaree SR, M. M. Sadeeq, S. F. Kak, H. S. Yahia, M. R. Mahmood, *et al*., "Comprehensive survey of big data mining approaches in cloud systems," *Qubahan Academic Journal*. 2021;129–38.
35. Amanuel SVA, Ameen SYA. "Device-to-device communication for 5G security A Review," *Journal of Information Technology and Informatics*. 2021; 126–31.

36. Bagui S, X. Wang, S. Bagui, "Machine Learning Based Intrusion Detection for IoT Botnet," *International Journal of Machine Learning and Computing.* 2021; 11.
37. Based Nature-Inspired Algorithms Optimization Scheduling," *Asian Journal of Research in Computer Science.* 2021;1–16.
38. Rawat, R., Rajawat, A. S., Mahor, V., Shaw, R. N., & Ghosh, A. (2021). Surveillance Robot in Cyber Intelligence for Vulnerability Detection. In *Machine Learning for Robotics Applications* (pp. 107–123). Springer, Singapore.
39. Yahia HS, Zeebaree SR, Sadeeq MA, Salim NO, Kak SF, A.-Z. Adel, *et al.*, "Comprehensive Survey for Cloud Computing
40. Sadeeq MA, Zeebaree S. "Energy management for internet of things via distributed systems," *Journal of Applied Science and Technology Trends.* 2021;259–71.
41. Yasin HM, Zeebaree SR, M. A. Sadeeq, S. Y. Ameen, I. M. Ibrahim, R. R. Zebari, *et al.*, "IoT and ICT based Smart Water Management, Monitoring and Controlling System A Review," *Asian Journal of Research in Computer Science.* 2021;42–56.
42. Rawat, R., Mahor, V., Chirgaiya, S., & Garg, B. (2021). Artificial Cyber Espionage Based Protection of Technological Enabled Automated Cities Infrastructure by Dark Web Cyber Offender. In *Intelligence of Things: AI-IoT Based Critical-Applications and Innovations* (pp. 167–188). Springer, Cham.
43. Rawat, R., Garg, B., Mahor, V., Chouhan, M., Pachlasiya, K., & Telang, S. Cyber Threat Exploitation and Growth during COVID-19 Times. In *Advanced Smart Computing Technologies in Cybersecurity and Forensics* (pp. 85–101). CRC Press.
44. Mahor, V., Rawat, R., Kumar, A., Chouhan, M., Shaw, R. N., & Ghosh, A. (2021, September). Cyber Warfare Threat Categorization on CPS by Dark Web Terrorist. In *2021 IEEE 4th International Conference on Computing, Power and Communication Technologies (GUCON)* (pp. 1–6). IEEE.

14

Applying Nature-Inspired Algorithms for Threat Modeling in Autonomous Vehicles

Manas Kumar Yogi*, Siva Satya Prasad Pennada, Sreeja Devisetti and Sri Siva Lakshmana Reddy Dwarampudi

Department of Computer Science Engineering, Pragati Engineering College (Autonomous), Surampalem, A.P., India

Abstract

This paper focuses on the security properties of autonomous vehicles and the inherent threat posed by the limitations of design principles applied as part of traditional smart vehicle design. Our paper sheds light on the threat models and adversary models where knowledge of the attacker plays an important part in analyzing their attack patterns and how the attacks develop over time. For the most part, nature-inspired algorithms consider optimization problems as a black box. Therefore, it isn't important to compute subordinates of the inquiry space. This reality makes nature-inspired algorithms profoundly adaptable with regard to taking care of different sorts of problems. Our paper applies nature-inspired algorithms to optimize the threat analysis model and this helps to develop measures to mitigate them. The nature-inspired algorithms have a good convergence rate to find an optimal solution so this model can be applied to develop a threat model suitable for autonomous vehicles. Attacker-based threat models are the strongest class of threat model, so we introduce the nature-inspired algorithms in this aspect of threat modeling. We develop design guidelines for enhancing the security in autonomous vehicles and also show the experimental results proving the better performance of nature-inspired algorithms for threat modeling in autonomous vehicles when compared to other popular methods. We also highlight a few future directions in this regard which will encourage researchers working in the design area of autonomous vehicles. We conclude the paper by advocating robust principles in design and development of autonomous vehicles.

Keywords: Autonomous vehicles, attacker, security, sensors, evolutionary algorithms, nature-inspired algorithms, threat models, trust

**Corresponding author*: manas.yogl@gmail.com

Romil Rawat, A. Mary Sowjanya, Syed Imran Patel, Varshali Jaiswal, Imran Khan and Allam Balaram (eds.) Autonomous Vehicles Volume 1: Using Machine Intelligence, (253–276) © 2023 Scrivener Publishing LLC

14.1 Introduction

An Autonomous vehicle (AV) is a driverless car that can sense the surroundings and drive according to it without any intervention from humans. Any weakness in the autonomous vehicle gives rise to a threat which is, in turn, a risk to the car as well as the passengers in it [1]. It is not only a faster means of travel but also fuel-efficient so it causes less harm to the environment. They are also used in mining and have seen significant results as they eliminate dangers to humans such as respiratory problems, injuries, etc. AV has undergone many transformations and it still has more to do in terms of security and performance [2]. Over the years, many companies have come forward with the idea of improving and deploying autonomous vehicles on the road so that accidents are reduced and passengers feel safer.

But still we need to analyze and remove weaknesses that can cause any sort of threat. (A threat is the thing that can damage or cause harm to the system.) There are still threats to passengers' privacy as there are malicious ways that an unauthorized organization or a person could hack the information from the database and use it to cause damage. AVs receive a lot of data from the sensors to make accurate predictions based on the environment. So, an AV is packed with a lot of sensors which could be the main source of threat. Threats could be to the hardware part due to any system failure or mode failures or any other architecture problem. It could be in the software part utilizing design or any assumptions taken while designing an algorithm for software deployment [3]. It could also be in the transmission medium between hardware and software parts. In this paper, we determine the extent of control on AVs, the system's security aspects, and ways to improve on areas where there is a chance of threat as security is as important as functionality.

An autonomous vehicle is a vehicle fit for satisfying the fundamental transportation abilities of a customary vehicle. The primary distinction to a conventional vehicle is a Driving Automation System (DAS) intended for AV [4]. DAS gives driving mechanization to the vehicle stage, along these lines offering the chance of generally changing transportation to lessen crashes, energy utilization, contamination, and cost of blockage. Such a vehicle draws much consideration from the scholarly world, industry and government. AV is a security basic framework. Any flaw in AV might bring about serious human injuries or even death. In the meantime, as a cyber-physical framework, an independent vehicle comprises a horde of heterogeneous parts, both cyber and physical, which represent extra security challenges. The complicated collaborations between these parts inside the AV make it hard to show the framework, and to adjust the well-being and security in an independent vehicle. Adjusting well-being and security is pivotal for independent vehicles, since any disappointments or assaults might prompt security misfortunes. This won't fulfill the developing demand on driving robotization frameworks. Distinctive degree of DAS is relating to various driving capacities and security necessities. Likewise, various levels will confront more possible hazards, threats, and difficulties. Hence, it is important to consider DAS when we dissect well-being and security for AV framework, on the grounds that the choice of security and security countermeasures for an AV with a similar driving capacity varies depending upon its robotization level. Nonetheless, ISO 26262 does not think about driving mechanization levels and accepts that a human driver is consistently present [5]. SAE J3061 is a cyber-security manual for vehicle frameworks, which characterizes the life cycle interaction system, and gives core values and so forth. In SAE J3061 [6], the cyber-security life cycle can be partitioned into a few stages: idea stage, item improvement stage (framework level, equipment level and programming level), creation and activity stage. The idea stage is the initial step for the entire life cycle, which incorporates the accompanying exercises: highlight definition, threat analysis and risk assessment, useful security idea, security necessities, and security assessment. The element definition characterizes the framework being created, to which the cyber security interaction will be applied, i.e., it characterizes the limit of the elements.

Threat Analysis and Risk Assessment (TARA) recognizes threats and evaluates the risk, and the after-effect of TARA drives generally downstream actuates. Security idea depicts the general methodology for getting security from TARA stage, and when not really settled for fulfilling the component, the security necessity is still up in the air. Security assessment is performed to recognize the current security stance of the cyber-physical vehicle, and it is created in stages all through the security life cycle. ISO 26262 is a global standard for practical well-being of E/E frameworks underway cars characterized by the International Organization for Standardization, which gives a car security life cycle (the board, improvement, creation, activity, administration, decommissioning) and upholds fitting the vital exercises during these stages. In the advancement part, correspondingly to SAE J3061, the well-being system is made out of Hazard Analysis and Risk Assessment (HARA) [7], practical security idea, security necessity and well-being assessment.

The situation where an independent vehicle is a common asset (vehicle sharing, for instance), makes actual admittance to the vehicle almost certain. Subsequently, we will play out an investigation on the interactor outside substances and its information stores, the two-way information stream between outside elements and the detecting, the infotainment framework process just as its two-way information stream from/to the door process [8]. Since the information stream from the detecting to the infotainment framework process is an active one and started from the interior trust limit of a vehicle. The information stream between outside substances and outside elements' information stores is likewise not thought of, since it is accepted that both of these parts are privately associated with one another and in this manner they are in one trust zone.

1. Interactors: According to scientists, interactors are simply helpless to ridiculing and renouncement. For our situation, vehicles, people on foot, clouds and framework are inclined to these sorts of assaults. In a parodying assault, the aggressor could take on the appearance of one of these elements, sending bogus data to our vehicle. As an illustration the aggressor could claim to be an alternate vehicle sending mistaken area data or profess to be the foundation cautioning the casualty's vehicle about a phony blockage to control the casualty vehicle's course. Another model is an assailant professing to be a passerby on the way to carry the vehicle to an undesired end. Concerning renouncement, an assailant could partake in a communication with the casualty's vehicle and later cases that this communication never occurred. For example, a malevolent vehicle following the casualty's vehicle conveys bogus data about speed and surmised area to the casualty's vehicle to control the casualty's vehicle's course [9]. Later on, the casualty's vehicle has no real way to recognize the offending vehicle. This sort of assault is regularly joined with the satirizing assault.
2. Data Store: Data stores are inclined to mess with information, data revelation and DoS. We accept that the information store of the external substances isn't remote, yet it is privately executed. Right off the bat, information saved in the information store can be altered, for instance through SQL infusion or through another weakness in the external substance which permits admittance to the information store [10]. An assailant could, for example, control the world model of the foundation element which then, at that point, sends bogus data to various vehicles. Besides, data revelation implies that information, for example, individual data of the

clients, can be perused by an aggressor. Outside element, for example, clouds and vehicles may hold client security information like the proprietor's name, plate number, vehicle speed and driving course which ought to be concealed from unapproved parties. Finally, information stores additionally face the danger of DoS. Since we accept that these information stores are privately carried out and are not open from the Internet, they are not less inclined to Distributed Denial of Service (DDoS) assaults [11]. A model would be a cloud and information store engineering where the information store is simply available to the cloud server in a similar LAN. DoS can be accomplished by bringing the information stores' servers down with some sort of code infusion, for example, SQL infusion.

3. Processes: Infotainment framework process is powerless to each of the six danger classifications. As per a few analysts, in term of processes, ridiculing personality and altering information allude to alteration or substitution of the process parallel. These assaults are frequently executed together. For example, to parody personality of another client an assailant needs to mess with the process double. These assaults can likewise prompt DoS by adulterating the memory information, or lead to height of advantage by messing with the memory. As may be obvious, these assaults are regularly caught with every others and occur as follow-up outcomes. Along these lines, DoS and height of advantage will be classed as a subsequent outcome; however, not as an independent danger. A model for these sorts of assault is Jeep Hack [12]. The analyst had the option to supersede (altering and caricaturing) the firmware of the CAN Bus regulator [13] which the infotainment framework is associated with by infusing the code through the cell or Wi-Fi organization. Accordingly, the specialists had unlimited authority over the vehicle functionalities. Data exposure for processes can mean separating information from the memory itself. This frequently can be accomplished with support flood. This data could be program code giving the aggressor more data on the most proficient method to take advantage of the framework or individual data like GPS area. Disavowal danger implies it is feasible to make a move and later on conceivably deny having taken it. In this particular case, it may

include a process adjusting information/code/twofold and having the option to guarantee that it didn't do it [14]. This process is of a second-level looking at to messing with information, since the danger of disavowal from pariah element will be limited if altering information danger is wiped out.

4. Data Flow: According to observation, information streams are presented to messing with information, data divulgence and DoS [15]. First we start with the information stream between outside elements and the detecting. The information traded between these two parts incorporates: (1) individual data identified with the vehicle such as vehicle's ID, area, plate number and (2) media information like music, video, news and so on. Information altering is conceivable.

An assailant can go about as man in the center and control the information between the two gatherings giving bogus data to both outside substances and the vehicle accomplishing comparative objective as satirizing assaults on external elements clarified previously. Data divulgence is another danger. Like messing with information, aggressors can dispatch a listening in assault to get to the information traded between the two players referenced previously. This information, particularly close to home data, can be utilized to recognize the proprietor of a vehicle or abused to follow the vehicle's area [16]. DoS danger keeps the two players from conveying with one another. This could be accomplished for instance with a man in the center assault dropping all communication between the two players. Second, the information stream between infotainment system process and the entryway is powerless to similar dangers the past information stream faces. The information trade between these two parts incorporates:

1. Data about the vehicle's state from any remaining ECUs, for example, power-train, vehicle security, solace and telematics to be shown for vehicle's clients.
2. Control signals from infotainment framework to non-security basic ECUs, for example, solace ECUs [17]. Since the communication between the infotainment framework process and the entryway takes place locally inside the vehicle engineering, assaults on the information stream between these two parts require actual access. However, an assault, for example, data exposure can uncover profoundly private data about the vehicle's state. An altering assault can hypothetically accomplish a similar goal like the Jeep hack

referenced in including a DoS assault which can obstruct the communication between infotainment framework process and the door which will forestall vehicle's client to get to data about the vehicle state and control specific vehicle's solace framework through the infotainment framework. This can be accomplished, for instance, by altering the process two-fold to drop all messages sent between the two parts.

The main sensors used in AVs are LIDAR (Light detection and ranging) [2, 18] for land surveying and making high-resolution maps, RADAR (radio detection and ranging) that uses radio waves to determine the distance, angle, or velocity of objects and cameras. GPS (Global Positioning System) is used in AV to locate any object's (obstacle) exact position. Ultrasonic sensors work with the help of ultrasonic sound waves to detect objects when they are very close to the AV. These sensors are placed at the right and left sides of the AV and are very helpful in places like car parking, etc. Odometer sensors are used to analyze the distance traveled by the vehicle. These are some main sensors that help in the determination of various factors and the functioning of AV. Have you ever wondered how vehicles communicate and make decisions according to their environment? Now let us discuss how communication happens in Autonomous vehicles.

C - V2X communication stands for Cellular Vehicle-to-everything communication [3, 19] which is mainly helpful in providing a 360 vision to the vehicle by taking the signals from the nearby vehicle. This helps in preventing road accidents and provides better transport.

The types of C - V2X communication are:

1. Vehicle to vehicle communication (V2V):
   In this communication, the vehicles send signals to nearby vehicles about their breaking speed, the direction of travel, and nearby threats to it which help in better communication between the vehicles. When the vehicle applies brakes, the nearby vehicles get informed and react according to the environment, like slowing down the vehicle.
2. Vehicle to infrastructure (V2I):
   In this communication, the data sharing is between the vehicles and their components. Some components involved here are traffic lights, video cameras, and lidars. If the traffic light is shown red, it is identified by the cameras of the AV, it decides to slow down the vehicle, and finally stops the vehicle.

3. Vehicle to pedestrian (V2P):
   It involves direct communication between the vehicle and pedestrians. When the vehicle comes closer to the pedestrian, the signals will be sent to both and are alerted. Here, communication is done by using network devices.
4. Vehicle to network (V2N):
   This helps in long-range communication using the cellular network for V2X communication. This helps in informing the nearby threats to other vehicles even if they are at long ranges.

The main focus of this paper is to identify threats and mitigate them. So, there are mainly three stages in the identification of threats to AV.

Threat identification:
In this stage, we will identify the threats which can be present in the vehicle system. Here, we will analyze what can go wrong in the system which acts as vulnerabilities (a weakness in design or implementation). Vulnerabilities in the system can happen due to weakness in the design, not enough assumptions while designing an algorithm, mode failure, or weakness in the communication [20].

Threat analysis:
It is a process which helps to find insecure components and which components of the system are at security risk and should need to be protected. Some possible areas where the threats can happen are Server, network ports, algorithms and software [21].

The possible threats are unauthorized access to the car database, Spoofing the GPS signals, denial of request from server (DDOS attack), damage of OS, software, improper functioning of components of the vehicle, tampering with the data, physical damage of the vehicle.

Threat modeling:
It is a process by which we can identify the potential threats in an insecure system and vulnerabilities which are caused due to mistakes in the system architecture. Various methodologies can be used to identify the threats.

The idea of self-driving vehicles started in the early 1920s with a transmitting antenna. It was operated by a person in some other car that sent radio impulses and these impulses were caught by the antenna. From then trials are being constantly made such as operating via electromagnetic fields, wires in the vehicle, electronic devices, and magnetic cables present in the roadway buried cables, etc. Research began in the way of the intelligent automated logic required for AV at the Univer Coordinated Scientific Laboratory [22]. The ALV (Autonomous Land Vehicle) project conducted the first demonstration of road tracking using lidar, computer vision, and autonomous robotic control in the 1980s. The ParkShuttle is the world's first AV that used artificial reference points recessed on the surface of the road to check its position [23].

As the years have passed, many companies have made significant improvements in AVs but there are still several threats AVs are facing. In July 2015, Google's test vehicle had several accidents which were caused by other cars in which there were human drivers. On February 14, 2016, a Google vehicle tried to avoid sandbags blocking its trajectory while maneuvering, and it collided with a bus [24]. If it hadn't moved, the collision would not have happened. On May 7, 2016, Tesla's electric car did not set the brakes while Trailer made a left turn and had a crash with a Tractor Trailer. On November 9, 2017, an auto-bus with passengers was involved in an accident with a truck [6, 25]. The truck was found to be defective in the accident, reversing on the stationary automated bus. The automated bus did not take evasive action or apply defensive driving techniques such as flashing lights or honking. Uber's autonomous car struck and killed a 49-year-old woman in Tempe, Arizona, by not triggering emergency stops in hazardous situations [27]. In 2018, in the Tesla cars, lidar was attacked due to vulnerabilities found in the insecure wireless communication channels and wireless-enabled onboard unit vulnerabilities helped attackers to prevent the vehicle from getting sensitive information using jamming. This vulnerability was fixed by assigning IPs to vehicles during the message transfer.

In 2018, in the BMW cars, due to vulnerabilities found in BMW OBD-2, hackers inserted the malware in the OBD2 and the system which made the vehicle operations fail [28, 29]. Later, BMW fixed the vulnerabilities in the OBD and TCU in their OS update.

In 2019, many companies faced an attack on GPS due to vulnerabilities found in the GPS location prediction. Using spoofing attack, the attackers changed the GPS functioning by reading it at some other location, like 9-10 meters other than the actual location, which caused the vehicle to make wrong decisions like stopping or moving in the wrong direction. This threat was fixed by using an antenna which was used to protect against interference and block the spoofing signals [30, 31].

All these incidents show that even though a lot of improvements have been constantly made there are still several threats to AVs due to system failure, communication transmission failures, etc. Some of the main obstacles for AVs are that they cannot run in chaotic environments, cars' sensing and navigation could fail in unpredictable environments, large animals and pedestrians couldn't be detected properly, road infrastructure is lacking required technology, failure to alert the driver or disengaging itself while alerting, failure to detect lane markings, etc. To strengthen the security system of AVs, they are equipped with VANETs. But some of the features of VANETs make them more vulnerable to threats and vulnerabilities at all their communication layers and make them insecure and cause issues like high dynamic topology, speed, and mobility of the car, etc. [32].

The components which can be attacked are sensors like lidars, radars by attacking the blind spot and showing fake images, spoofing of GPS signals [33], introducing malware in software which results in stealing the user's data, improper functioning of components.

Some types of attacks which are mainly performed in the AVs are:

A) DOS attack (denial of service):
DOS attack, which is also known as the distributed denial of service attack, is one of the most dangerous attacks on self-driving cars. This attack is done on the components like lidars, radar, cameras, braking systems which result in improper functioning and lead to accidents (damage to the vehicle).

B) SQL Injection:
The SQL language is used to store and retrieve the data from the database. The car contains a database that helps in the functionality of the components.

SQL injection is a type of cyber-attack in which the attacker uses a SQL code to manipulate the database and gain access to the required information.

C) Spoofing of signals:
Spoofing is a cyber-attack in which the attacker tricks the computer network by using false identities to gain access to the vehicle. The spoofing can be performed on sensors like GPS, lidars, etc.

The attacker tricks the lidars by sending the signals for a very short time and makes it create an image that is not present, and the GPS can be spoofed by sending the false signals to the target receiver and gradually increasing the signals so that original signals can be drawn away from the receiver, so that the estimated position in the receiver is different from the actual location.

D) Man-in-the-middle attack:
AVs use wireless communication to communicate with the nearby vehicles, infrastructure, and network. In the V2V communication, the attacker

manipulates the messages which are shared between two vehicles while the vehicle believes that they are in direct communication with each other so that the false information can be shared between them.

14.2 Related Work

As of now, homegrown and unfamiliar researchers have set up an assortment of cybersecurity danger investigation structures; however, the examination cycle is profoundly emotional and needs quantitative examination. Formal and quantitative TARA approaches are an exploration heading that can adequately tackle this issue. Formal quantitative danger examination technique utilizes normalized dialects like SysML to officially depict the framework under test and lead danger demonstrating at the framework level [34]. In expansion, through conventional displaying, probabilistic investigation of vehicle framework network security can be accomplished, in this way accomplishing more itemized quantitative TARA.

TARA Methods [4, 35] with Trade-Off Considerations: The expanding intuitiveness among digital and vehicle frameworks and network bring about new well-being and security challenges.

Since cybersecurity assaults can influence the useful well-being of vehicles, it is unreasonable to reinforce the general protection level without one or the other side. Also, an excessive number of security protection systems not exclusively will expand the general vehicle cost, yet may even influence the client experience. Along these lines, considering the compromises of security, well-being, vehicle cost, and client experience is a significant heading of TARA strategies.

Information Driven TARA Process: As present-day vehicles progressively trade information with the cloud, OEMs can gather all the more genuine information from clients' vehicles. A lot of information can bring numerous opportunities for TARA. For instance, the TARA interaction dependent on AI calculations has extremely high prerequisites for information greatness. Enormous scope information can give an assurance to the precision of danger model training. The information driven TARA process is another examination bearing.

The game-theoretic methodology [36] joins attack-guard tree with game hypothesis. Game hypothesis is an investigation of the numerical model of vital cooperation among normal leaders. The game-theoretic methodology can give inside and outside information on the procedures embraced by attackers and protectors. As per the attack-safeguard tree, the attacker has a few attack techniques to accomplish the attack objective, and each attack strategy might relate to a few countermeasures. The importance of the game hypothesis is to assist safeguards with picking the best alleviation and expand their result. Initial, an attack-safeguard tree ought to be set up, so every one of the possible attacks and alleviations can be recorded. Then, at that point, by applying game hypothesis on the tree, the protector can arrive at ideal alleviation, which is firmly identified with the attack procedures. In any case, the game theoretic methodology is established on the way that the players act reasonably, which once in a while is absurd actually. In addition, the utility capacity should be appropriately planned. OMNeT++ is an open-source, particular, part based C++ reenactment library and structure that can be utilized to reenact vehicle organizations [37]. OMNeT++ can undoubtedly construct network models and has high reenactment granularity. Furthermore, it can likewise perform network attack reenactment and threat investigation. The information recording capacity can mirror the effect of various sorts of attacks on the information in the organization. Specialists have utilized this device to set up a Sybil attack model against vehicle lines and assessed the effect of the attack on vehicle network execution. Down to earth Threat Analysis is an apparatus that can be utilized for threat demonstrating and programmed computation of hazard appraisal results. From the outset, it needs to set different boundaries like framework resources, threats, taken advantage of weaknesses, comparing relief measures, attack types. The threat model is put away in a powerful data set so the model boundaries can uphold dynamic changes. By constantly overhauling the boundaries of the model, it can guarantee that the hazard evaluation and security the board cycle can be done ceaselessly and viably. It develops a particular threat situation to show the weaknesses that a certain threat can use to attack the resources of the framework. At a similar time, countermeasures for the threat ought to be added.

The EVITA strategy is a resource-based threat analysis technique [38]. This strategy gives a savvy security engineering that can give far-reaching security in various advancement stages like plan, check, and model for vehicle organizations. The EVITA technique plays out an attack appraisal for every resource in the framework and afterward surveys the degree of risk that the attack might cause. Risk is an element of the attack probability and the seriousness of the mischief brought about by the attack. In view of these, the threats are Risk-evaluated and the threat is still up in the air. The EVITA risk appraisal technique can be applied to survey likely threats. The distinguished potential threats can be positioned by the risk level to additional attention the analysis on the most elevated risk threats. Then, at that point, the organization security objectives not really settled for the most elevated risk threats. Notwithstanding, the EVITA strategy just gives an assessment technique and doesn't give a total assessment process, which will carry inconvenience to clients. HEAVENS strategy makes up for this imperfection [39]. The mix of security targets and level of effect during threat analysis assists with evaluating the potential business effect of a threat on important partners. HEAVENS is, thusly, an entirely reasonable appraisal strategy for assessing the data security risks of car electronic and electrical frameworks. Simultaneously, the HEAVENS technique gives an itemized interaction of threat analysis and risk appraisal, which significantly diminishes the trouble of utilization and expands the attainability of the strategy, which is additionally an essential for its inescapable use.

The BRA (Binary Risk Analysis) surveys the resources for be ensured in the framework by carrying out a cycle [39, 40]. The BRA strategy can be utilized for fast risk discussions to talk about explicit risks in only a couple of moments. All things considered, the subsequent risks are just named high, medium, or low. Moreover, a moderate analysis pattern prompts threat order exclusively of high risks. Also, no organized assessment of threat situations is given, and the subsequent threat order is excessively simple for idea advancement stages. Safeguard is a multi-metric way to deal with assesses the framework's degree of security, protection, and steadfastness. The principle objective of this technique is to assess different framework designs and select those that meet or accomplish set up prerequisites. In the NHTSA approach [40], all important locally available parts and frameworks have been thought of, and the information stream and the trust limit between the parts can be outwardly noticed.

14.3 Proposed Mechanism

In our work we use a certain branch of nature-inspired programming to model the threats and perform threat analysis on autonomous vehicles. Specifically we select the principles of evolutionary programming to develop a threat model from an adversary point of view. In the 1960s the concept of Evolutionary programming (EP) was advocated by Lawrence Fogel, Al Owens and Jack Walsh. An EP does not involve recombination but rather evolves a population of individuals. New individuals are created only by mutation. EP is suitable for our case due to the infinite number of states in which the attacker may bring the vehicle under control. Some of these states may be highly undesirable, which may even cause harm to a user. There are variants of EP which can be applied to develop a threat model and perform risk analysis in the context of autonomous vehicles.

Threat model formulation using the continuous evolutionary programming: Suppose we want to minimize the adversary threat value f(x), where x is an n-dimensional vector. Assume that f(x) > 0 for all x. An EP begins with a randomly-generated population of individuals $\{x_i\}$, i £ [0, n]. We create children $\{ x'_i \}$ as shown below:

$$x'_i = x_i + r_i(\beta\ f(x_i) + \Omega)^{1/2} \text{ where i=1 to n} \qquad (14.1)$$

Where r_i represents n-data vector which is random, each of whose data is extracted from a Gaussian distribution with a mean of 0 and a variance of 1, and ß and Ω are EP tuning factors. The variance of the mutation of X{ is (ßf(xi) + 7). If ß = 0, then all individuals have the same average mutation magnitude. If ß > 0, then an individual with a low cost does not mutate as much as an individual with a high cost. Often ß = 1 and Ω = 0 are used as default, standard EP parameter values.

Our threat model tries to minimize the f(x) value so that even if the values of individuals increase or decrease, it does not affect the threat function value. Once we find the lower bound of the threat value, the adversary cannot affect the working state of any entity in the autonomous vehicle. The graphical result shown in Figure 14.1 depicts the relation between the tuning factors, ß and Ω against the percentage of attack severity in our proposed threat model.

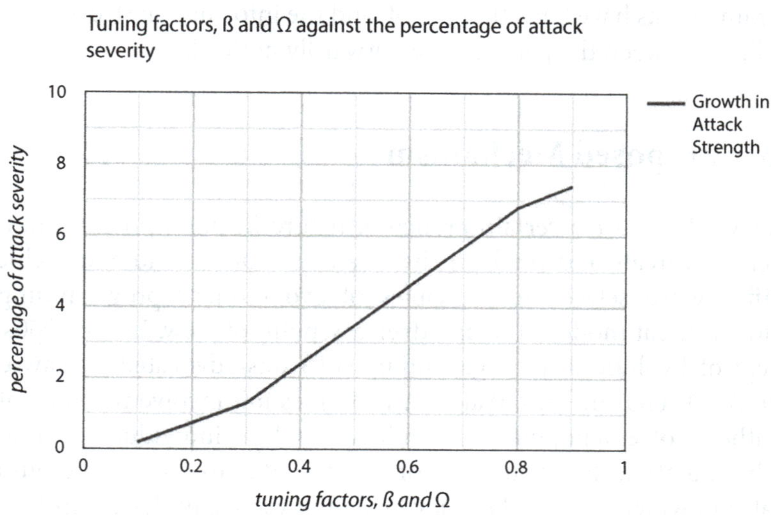

Figure 14.1 Effect of tuning factors against the percentage of attack severity.

Threat model based on discrete evolutionary programming:
In this type of threat model we use the "random mutation" for FSM optimization selected randomly as one of the following.

We assume the adversary has control over certain units of autonomous vehicle and has the following abilities:

- He can Add a state with random input/output and input/transition pairs.
- He can Delete a state. Any state transitions to the deleted state are redirected to

Another randomly selected state.

- He can randomly change an input/output pair for a randomly selected state.
- He can randomly change an input/transition pair for a randomly selected state.
- Randomly change the initial state.

To refine the threat model, we can think of introducing a penalty factor also which may reduce the degree of threat based on the choice of the states which can be as random as possible. This partiality towards the selection of the best N individuals at the termination of each generation to simpler state machines. This logic helps us to find FSMs to produce a desired pattern, and also to determine unique FSMs to deduce a required pattern.

The below logic summarizes the principle used in our proposed threat model:

{xi} <— {randomly generated population}, where i to 1 to n

While not (termination criterion)
Calculate the cost $f(x_i)$ of each individual in the population
For each individual x_i, where i to 1 to n

x_i' <— random mutation of x_i

Next individual

{ x_i } <— best N individuals from (x_i, x_i')

Next generation
In the above, we want to minimise the value of f (x_i), so that the threat degree is also reduced.

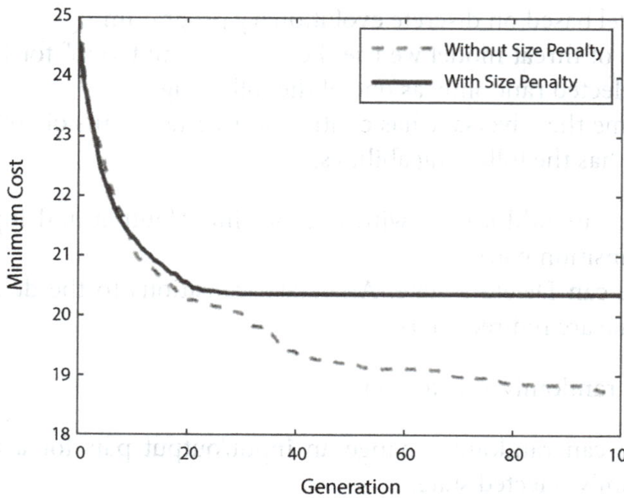

Figure 14.2 Generation of threats against minimum cost related to threats.

In Figure 14.2, we can observe that as the generation of threats increases, the minimum cost related to threats decreases with or without the size penalty. But when we compare with respect to the penalty factor then we can easily observe that the minimum cost decreases with a higher rate when compared to minimum cost with size penalty.

14.4 Performance Results

From the result as shown in Figure 14.3, we observe that all the threat models have increasing nature with respect to increase in nature of threats and attack severity. We find that our proposed method of threat model based on continuous evolutionary programming has the highest percentage of attack severity detection rate as number of threats increases. It is nearly 2% better than the most popular method, i.e., attack defense tree approach.

Now, for the performance of our second proposed method, as we can observe from Figure 14.4, all the threat models have increasing nature with respect to increase in nature of threats and attack severity. We can determine that the threat model based on discrete evolutionary programming [26] performs better than other popular methods. This is due to the fact that it selects the states of the system in a random order due to which it is difficult to predict the threat severity of an attack.

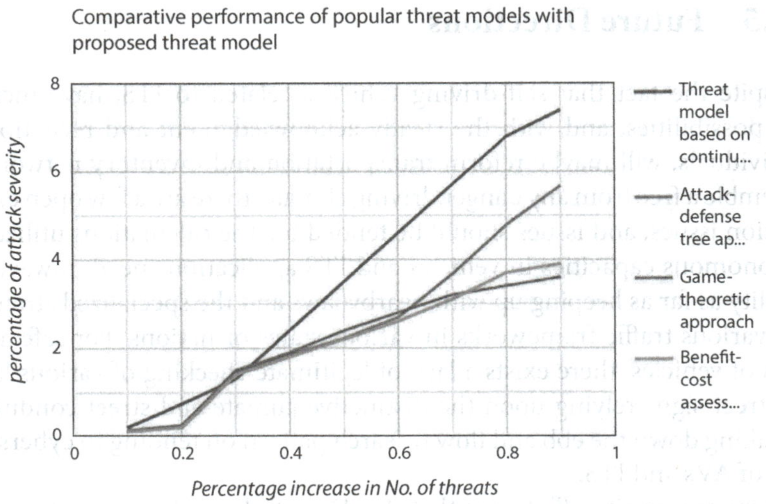

Figure 14.3 Comparative performance of popular threat models with threat model based on continuous evolutionary programming.

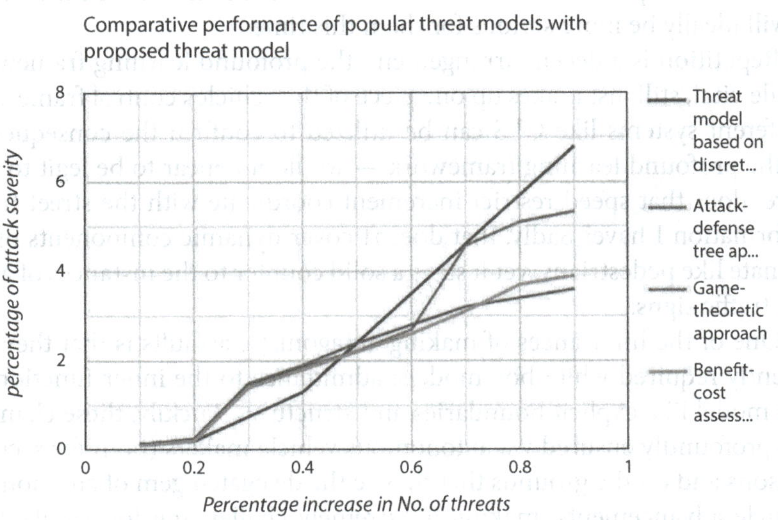

Figure 14.4 Comparative performance of popular threat models with threat model based on discrete evolutionary programming.

14.5 Future Directions

Despite the fact that self-driving vehicles, related to ITS, have incredible possibilities, and with the steady acknowledgment and reception by individuals, will maybe reform transportation and inventory network, to assemble a free from any danger driving climate, there are a few open examination issues, and issues should be tended to. The calculations utilized in autonomous capacities in vehicles and ITS applications need unwavering quality as far as keeping up with nearby laws and the specialized strategies for various traffic frameworks in various states or nations. For safe activities of vehicles, there exists a lack of legitimate checking of various kinds of street signs relying upon the distinctive climate and street conditions. Breaking down the ebb and flow research pattern on tending to cybersecurity of AVs and ITS.

Anyway, on the off chance that the AV is to have the option to peruse a stop sign accurately, what should be possible? Eliminating the bothers to all possible signs is a choice, *yet all* the same that is a significant overhead. A more sensible methodology is to foster machine learning systems that are strong against these sorts of assaults, mostly via preparing them to effectively distinguish ill-disposed models — an openness treatment of sorts. Unfortunately, our wily assailants could simply make another antagonistic technique that the framework has not seen, in spite of the fact that it will ideally be more serious for them this time.

Repetition is a decent arrangement; the profound learning framework, while vital, still just makes up one piece of the vehicle's control framework. Different systems like GPS can be utilized to confirm the consequences of the profound learning framework — would it appear to be legit to stop here; does that speed restrict increment coordinate with the street traffic information I have? Sadly, that doesn't cover dynamic components of the climate like pedestrians, yet it stays a solid counter to the instances of altering traffic signs.

One of the hindrances of making antagonistic assaults is that they frequently required white-box models: admittance to the inner functions of the model like explicit boundaries and structures. Luckily, these elements are profoundly ensured via autonomous vehicle makers (both for security reasons and on the grounds that they're the delegated gem of autonomous vehicle advancement), making it extremely challenging for assailants to access to create their assaults. Notwithstanding, investigation into antagonistic assaults against discovery models is ongoing, so we're not exactly free and clear yet. Eventually, any strategies that add to expanding the power of the machine learning systems utilized in self-driving vehicles will be a welcome gift.

Assaults on Sensors: Sensors see the climate the AVs work in, and any assault on sensors would be inconvenient, compelling the vehicle to settle on off-base choices with genuine outcomes. The aggressors can focus on any sensor of an AV and attempt to fool the vehicle by giving tempered information. For instance, stop signs with added little spray painting or craftsmanship stickers could be difficult to be perceived by computer vision systems. Embedding counterfeit objects could compel the vehicle to stop or dial back superfluously, while erasing a genuine object could prompt mishaps. Mechanical progression makes it simple for the assailant to accomplish such treating with the sensor signals, particularly with Lidars sensors, as exhibited in, alongside the actual treating of street signals. New ways can be designed to dispatch hazardous assaults like crisis brake or AV freezing assault, where the vehicle can unexpectedly stop or remain stopped in a convergence even after the red sign becomes green. Object detection calculations should be adequately vigorous to identify continuously whether an object has been erased from the vision framework or any phony objects are embedded.

Remote controlling AVs: Another examination region that needs more consideration is the controller of an AV enduring an onslaught. At the point when an autonomous vehicle is enduring an onslaught, its tasks should be assigned to the drivers, or it tends to be controlled from a distance, which would be more fitting for cargo vehicles. The assurance of vehicle present (position and speed) is the way to control a vehicle from a distance, and arrangement of IoT sensors in ITS eventual the answer for vehicle present calculation. In any case, the information conveyed to the remote place through remote channels might actually be enduring an onslaught. Any such assault signals should be demonstrated so it tends to be separated from any direction. When identified, these assault signs could be stuck utilizing a procedure where a jammer(s), friendly to the AV and remote focus, will just obstruct assailants without hampering genuine correspondence.

Adversarial Machine Learning: Autonomous vehicles use machine learning calculations, particularly profound learning models, for decision making at various levels. Notwithstanding, ML procedures are helpless against painstakingly created adversarial annoyances. The blemished preparing process, the distinction in factual appropriations for preparing and activity information, the learning system being not really interpretable urge the enemies to assault the ML framework sent in AV. Despite the fact that adversarial assaults on object detection from picture have been concentrated comprehensively, something similar for Lidar point cloud and radar information have been less investigated, accordingly still stances digital danger to AV and need research consideration. Adversarial information age and model retraining strategies assume a significant part in making machine learning methods powerful and dependable. As of late, a retraining method with brilliantly chose adversarial tests has been proposed for recognizing malware in modern IoT applications with promising outcomes. Comparable methods can be investigated to configuration learning models profoundly strong against adversarial assaults in ITS and AVs.

Assaults on CAN: Controller Area Network (CAN) is a heritage framework that has been utilized as a correspondence channel between ECUs. More than 1,000,000 lines of programming code are prepared in an AV. All depend on CAN to work. Being an inheritance framework, it is constantly being presented to digital dangers, which should be tended to completely; in any case carrying out the most grounded cybersecurity for all subsystems in a vehicle would not ensure the vehicle being gotten from digital assaults. The CAN network needs ceaseless observing for any digital assault being recognized, and more examination should be done to make CAN as a safe correspondence media for AVs' subsystems. Machine learning-based models can be utilized to screen CAN to identify any abnormalities in signals communicated through CAN.

More extensive Scope of Cyber-assaults: The assault surface of an AV is moderately extremely enormous, considering a vehicle is an end result gathered from a large number provided by the Original Equipment Manufacturer (OEM). The manufacturers need to ensure each such part can battle cyber-attacks and furthermore ensure the correspondence between those parts is additionally gotten. Not at all like other cyber-systems where security comes after the usefulness, AV manufacturers have needed to interlace cybersecurity with usefulness that can possibly be feasible in case they make it an authoritative culture to accept the significance of cybersecurity. This would be one of the primary difficulties for AV to involve the street.

New advancements utilized in self-driving vehicles and ITS systems in various nations have not been entirely tried on streets with genuine traffic situations. As a result, the likely assaults and their potential outcomes are not obviously perceived by the applicable examination networks and partners.

Security assaults change their tendency as new applications and innovations are not liberated from weaknesses, and the new ones are found consistently. Countermeasure methods dependent on cryptography, access control, validation component, and actual layer equipment and correspondence security likewise keep on developing. In this powerful and invigorating examination scope, this paper presents excellent establishment information to the specialists from the scholarly world and industry.

14.6 Conclusion

This paper brings out the inherent strength of nature-inspired algorithms to design and analysis of threat models applicable for autonomous vehicles. Safety is the essential objective when planning autonomous vehicles. Any inadvertent disappointments (safety issues) or potentially deliberate assaults (security issues) for such vehicles might bring about serious safety misfortunes, e.g., human injury or even death. In this manner, the viable arrangement of safety and security for AVs is vital. The primary contrast among AV and ordinary vehicles is that there are various degrees of driving robotization in AV that characterize which functional and strategic capacities are performed by a human driver and by the driving computerization framework. In this manner, the determination of safety and security countermeasures for an AV with similar capacities contrasts relying upon its mechanization level. In this paper we have efficiently advocated a novel algorithm belonging to the category of evolutionary programming techniques which can be used to model threats to autonomous vehicles. Any attacker in an AV will definitely focus on attacking and compromising the safety measures available in the AV. So, we develop a threat model using the similar thinking of an adversary and how they will try to create abnormal behavior in the AV. Our experimental results show that our proposed methods are relatively better than existing popular methods used to develop threat models and in future work we plan to deploy these threat models to the recent datasets available in the open-source community to bring out any validity issues associated with our proposed models.

References

1. Y. Wang, Y. Wang, and H. Qin, "A Systematic Risk Assessment Framework of Automotive Cybersecurity," *Automotive Innovation*, vol. 4, pp. 1–9, 2021.
2. ISO, ISO/SAE FDIS 21434-2021, Road Vehicles—Cybersecurity Engineering, ISO, Geneva, Switzerland.
3. S. V. E. S. S. Committee, SAE J3061 Cybersecurity Guidebook for Cyber-Physical Automotive Systems, SAE Standard, Warrendale, PA, USA, Work-in-Progress, 2017.

4. O. Henniger, A. Ruddle, S. Herve, W. Benjemin, and M. Wolf, "Securing vehicular on-board it systems: the evita project," in *Proceedings of the VDI/VW Automotive Security Conference, Ingolstadt, Germany,* October 2009.
5. G. Macher, "A review of threat analysis and risk assessment methods in the automotive context," in *Proceedings of the International Conference on Computer Safety, Reliability, and Security,* September 2016.
6. A. Skavhaug, G. Jeremie, S. Erwin, and F. Bitsch, *Computer Safety, Reliability, and Security*, Springer, Berlin, Germany, 2016.
7. D. Dominic, M. E. Ryan, M. Di, and C. Sumeet, "Risk assessment for cooperative automated driving," in *Proceedings of the 2nd ACM Workshop on Cyber-Physical Systems Security and Privacy,* October 2016.
8. J. Dürrwang, K. Beckers, and R. Kriesten, "A lightweight threat analysis approach intertwining safety and security for the automotive domain," in *Proceedings of the International Conference on Computer Safety, Reliability, and Security,* September 2017.
9. M. Hagan, F. Siddiqui, and S. Sezer, "Policy-based security modelling and enforcement approach for emerging embedded architectures," in *Proceedings of the 2018 31st IEEE International System-On-Chip Conference (SOCC)*, pp. 84–89, IEEE, Arlington, VA, USA, September 2018.
10. F. Haidar, A. Kaiser, B. Lonc, and U. Pascal, "Risk analysis on C-its pseudonymity aspects," in *Proceedings of the 2019 10th IFIP International Conference on New Technologies, Mobility and Security (NTMS)*, pp. 1–5, IEEE, Canary Islands, Spain, June 2019.
11. J. Cui and G. Sabaliauskaite, "US2: an unified safety and security analysis method for autonomous vehicles," in *Proceedings of the Future of Information and Communication Conference, Springer, Singapore,* April 2018.
12. G. Macher, H. Sporer, R. Berlach, A. Eric, and K. Christian, "SAHARA: a security-aware hazard and risk analysis method," in *Proceedings of the 2015 Design, Automation & Test in Europe Conference & Exhibition (DATE)*, pp. 621–624, Springer, Grenoble, France, March 2015.
13. C. Schmittner, Z. Ma, and P. Smith, "FMVEA for safety and security analysis of intelligent and cooperative vehicles," in *Proceedings of the International Conference on Computer Safety, Reliability, and Security,* September 2014.
14. C. Schmittner, Z. Ma, S. Erwin, and G. Thomas, "A case study of fmvea and chassis as safety and security co-analysis method for automotive cyber-physical systems," in *Proceedings of the 1st ACM Workshop on Cyber-Physical System Security,* April 2015.
15. S. Verma, T. Gruber, P. Puschner, and C. Schmittner, "Combined approach for safety and security," in *Proceedings of the International Conference on Computer Safety, Reliability, and Security,* September 2019.
16. D. Suo and S. Sarma, "A test-driven approach for security designs of automated vehicles," in *Proceedings of the 2019 IEEE Intelligent Vehicles Symposium (IV)*, pp. 26–32, Springer, Paris, France, June 2019.

17. J. Cui and B. Zhang, "VeRA: a simplified security risk analysis method for autonomous vehicles," *IEEE Transactions on Vehicular Technology*, vol. 69, no. 10, Article ID 10494, 2020.
18. J. Monteuuis, B. Ayman, Z. Jun, H. Laiod, S. Alain, and U. Pascal, "SARA: security automotive risk analysis method," in *Proceedings of the 4th ACM Workshop on Cyber-Physical System Security, ACM, Incheon, Republic of Korea,* June 2018.
19. M. Zoppelt and R. T. Kolagari, "SAM: a security abstraction model for automotive software systems," in *Proceedings of the Security and Safety Interplay of Intelligent Software Systems, pp. 59–74, Springer, Barcelona, Spain,* September 2018.
20. S. Kim and R. Shrestha, "AUTOSAR embedded security in vehicles," in *Automotive Cyber Security,* Springer, Berlin, Germany, 2020.
21. T. Halabi, O. A. Wahab, R. A. Mallah, and M. Zulkernine, "Protecting the Internet of vehicles against advanced persistent threats: a bayesian Stackelberg game," *IEEE Transactions on Reliability*, IEEE, vol. 70, no. 3, pp. 1–16, 2021.
22. J. Yu and F. Luo, "A systematic approach for cybersecurity design of in-vehicle network systems with trade-off considerations," *Security and Communication Networks*, vol. 2020, Article ID 7169720, 14 pages, 2020.
23. S. Kim and R. Shrestha, "Internet of vehicles, vehicular social networks, and cybersecurity," in *Automotive Cyber Security*, Springer, Berlin, Germany, 2020.
24. J. Zhong, S. Du, Z. Lu, H. Xhu, C. Fan, and Q. Xue, "Security modeling and analysis on intra vehicular network," in *Proceedings of the 2017 IEEE 86th Vehicular Technology Conference (VTC-Fall)*, pp. 1–5, Toronto, ON, Canada, September 2017.
25. R. C. Rinaldo and D. Hutter, "Integrated analysis of safety and security hazards in automotive systems," in *Proceedings of the ESORICS 2020 International Workshops, CyberICPS, SECPRE, and ADIoT*, pp. 3–18, Springer, Guildford, U.K., September 2020.
26. P. Mundhenk, S. Sebastian, L. Martin, A. S. Fahmy, and C. Samarjit, "Security analysis of automotive architectures using probabilistic model checking," in *Proceedings of the 2015 52nd ACM/EDAC/IEEE Design Automation Conference (DAC)*, pp. 1–6, IEEE, San Francisco, CA, USA, June 2015.
27. D. Li, T. Liu, T. Cao, and P. Deng, "The risk assessment for unmanned vehicle using bayesian network," in *Proceedings of the International Conference on Geo-Informatics in Resource Management and Sustainable Ecosystem, Springer, Wuhan, China,* October 2016.
28. I. Friedberg, K. McLaughlin, P. Smith, D. Laverty, and S. Sezer, "STPA-SafeSec: safety and security analysis for cyber-physical systems," *Journal of Information Security and Applications*, vol. 34, pp. 183–196, 2017.
29. D. Ren, S. Du, and H. Zhu, "A novel attack tree based risk assessment approach for location privacy preservation in the VANETs," in *Proceedings of*

the 2011 IEEE International Conference on Communications (ICC), pp. 1–5, IEEE, Kyoto, Japan, June 2011.

30. M. Salfer, H. Schweppe, and C. Eckert, "Efficient attack forest construction for automotive on-board networks," in *Proceedings of the International Conference on Information Security*, October 2014.
31. M. Krisper *et al.*, "RISKEE: a risk-tree based method for assessing risk in cyber-security," in *Proceedings of the European Conference on Software Process Improvement*, September 2019.
32. N. Kaja, A. Shaout, and D. Ma, "Fuzzy based threat assessment model (FTAM)," in *Proceedings of the 2019 International Arab Conference on Information Technology (ACIT)*, pp. 144–149, Springer, Al Alin, December 2019.
33. W. Xiong, F. Krantz, and R. Lagerstrom, "Threat modeling and attack simulations of connected vehicles: proof of concept," in *Proceedings of the International Conference on Information Systems Security and Privacy*, Springer, Prague, Czech Republic, February 2019.
34. J. Santhosh and S. Sankaran, "Defending against Sybil attacks in vehicular platoons," in *Proceedings of the 2019 IEEE International Conference on Advanced Networks and Telecommunications Systems (ANTS)*, pp. 1–6, IEEE, Goa, India, December 2019.
35. S. Bahamou, J. Bonnin, and M. I. E. Ouadghiri, "Vehicular adhoc network's privacy assessment based on attack tree," in *Proceedings of the International Workshop on Communication Technologies for Vehicles, San Sebastián, Spain*, June 2016.
36. S. Garg and G. Aujla, "Accessing risk priority of SSL SYN attack using game theoretic attack defense tree model for VANETs," in *Proceedings of the 2016 3rd International Conference on Computing for Sustainable Global Development (INDIACom)*, pp. 729–734, Springer, New Delhi, India, March 2016.
37. S. Garg, G. S. Aujla, N. Kumar, and S. Batra, "Tree-based attack-defense model for risk assessment in multi-UAV networks," *IEEE Consumer Electronics Magazine*, vol. 8, no. 6, pp. 35–41, 2019.
38. S. Bahamou, D. E. Ouadghiri, and J. Bonnin, "When game theories meet security and privacy related risk assessment of vehicular networks (VANET)," *Journal Mobile Multimedia*, vol. 12, pp. 213–224, 2017.
39. A. Le and C. Maple, "A simplified approach for dynamic security risk management in connected and autonomous vehicles," in *Proceedings of the Living in the Internet of Things (IoT 2019)*, May 2019.
40. S. Longari, C. Andrea, M. Carminati, and Z. Stefano, "A secure-by-design framework for automotive on-board network risk analysis," in *Proceedings of the 2019 IEEE Vehicular Networking Conference (VNC)*, pp. 1–8, Springer, Los Angeles, CA, USA, December 2019.

15

The Smart City Based on AI and Infrastructure: A New Mobility Concepts and Realities

Vinod Mahor[1]*, Sadhna Bijrothiya[2], Rina Mishra[3], Romil Rawat[3] and Alpesh Soni[3]

[1]*Department of Computer Science & Engineering, IES College of Technology, Bhopal, M.P., India*
[2]*Department of Computer Science & Engineering, Maulana Azad National Institute of Technology, Bhopal, MP, India*
[3]*Department of Computer Science & Engineering, Shri Vaishnav Vidyapeeth Vishwavidyalaya, Indore, India*

Abstract

The reproduction of humans cognitive functions by technologies is known as Artificial intelligence (AI). AI is a powerful concept that, when used effectively, can aid long-term transformations towards more energy sources. The computational intelligence skills and abilities of artificial intelligence can be used as a strategy to enable computers to handle reconfigurable situations. The cityscape scenes that we have known for decades are leaving their place to a new age; the "smart city" has arrived. Transportation is among the most important industries where AI can make an impact. The adoption of smart transport systems in general and autonomous vehicle transport has the potential to significantly improve "mobility" and its impact on social development. Regardless of its technological perspective, this next-generation AI-based mobility must be a user "central technology" that "understands" and "satisfies" all. This paper covers smart city development of infrastructures and Mobility-based AI for smart homes. A creative approach can be used by exploring the underexplored intersection of AI and robots, transportation and the "smart city" and how this will affect the outcomes of cities. The chapter focuses on key integrated mobility projects such as Connected; Actions

**Corresponding author*: vinodengg.mt@gmail.com

Romil Rawat, A. Mary Sowjanya, Syed Imran Patel, Varshali Jaiswal, Imran Khan and Allam Balaram (eds.) Autonomous Vehicles Volume 1: Using Machine Intelligence, (277–296) © 2023 Scrivener Publishing LLC

that can function as Autonomous Vehicles (AVs), autonomy Personal Unmanned Aerial Vehicles (PUAVs) and Mobility as a Service (MaaS), as well as advanced transport and communication, e.g., Internet of Things (IoT) and Physical Internet (PI) or broader transformations, such as Industry 4.0

Keywords: Smart city, AI, infrastructure, mobility concepts, IoT and industry 4.0

15.1 Introduction

Artificial Intelligence has come about at a time when there is a greater necessity than ever to move to another technological social environment in the future to avoid the negative consequences of many sources and an unwitting proactive quality of life ideology of the future. The ability to act as a catalyst for change, Artificial intelligence (AI) is simply defined as the learning capacity of service. The capacity to analyze external realities appropriately learns from them and uses this learning to achieve certain goals and activities through adaptability [1, 2]. AI focuses on the creation of autonomous entities capable of reasoning and planning in the absence of an established knowledge base about their environment [3]. AI has the potential to make us healthier, richer, better by reducing our dependence on human labour and greatly increasing our scientific and technical progress [4, 5]. The Figure 15.1 represents framework relating to Application models of artificial intelligence and Figure 15.2 focussed on Analysis or different type research publication for Smart Industry with AI and Transportation for research reviews.

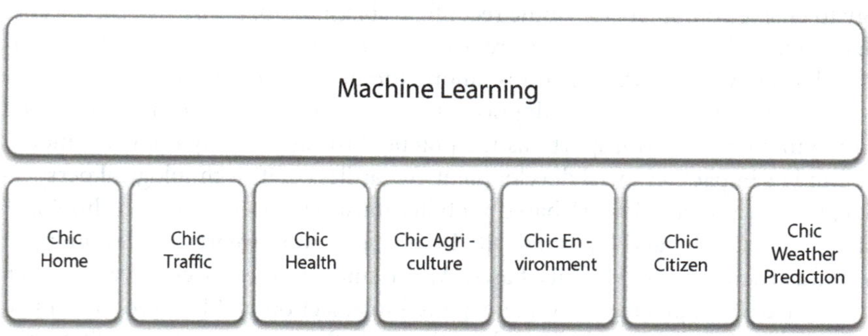

Figure 15.1 Application model of artificial intelligence.

AI has revolutionized our existence in a variety of ways, from semi-autonomous cars on the road to robotic vacuum cleaners in our homes, and it may be starting to permeate every part of our lives from health insurance to learning, fun, safety and in the near term [6–10]. One of the primary sectors where AI will have the biggest paradigm-shifting impact is transportation. This study provides a unique conceptual contribution through a description of another major important transportation component that is projected to be critical to the near future artificial intelligence-centric smart-cities. First, the essay functions as a study vocabulary, describing ideas and their dimensions in connection to the future mobility perspective that is autonomous, networked, sharing, and digital [20]. This paper focuses on intervention strategies such as "Integrated and Autonomous Vehicles" (CAVs), "Unmanned and Personal Aerial Vehicles" (UAVs and PAVs), and "Mobility-as-a-Service" (MaaS), as well as the "Internet of Things" (IoT), "Physical Internet" (PI), and Industry 4.0, which are three projects that may have a substantial onward journey and thus are critical elements [21].

Since public transportation is by far the most critical pillar of a city's operation, progress and profit, modernizing transportation changes the concept of the city; theories comprehend the relationship between road transportation and urban use. For example, the urban development rental model and the site view demonstrate that urban structure and transportation system development are inexorably intertwined [11]. The vision of the cities of tomorrow is inextricably linked with the future shape of transport networks [12]. The need for mobility solutions to combat overpopulation, pollution and environmental degradation is the foundation of what we call smart cities. A smart city has a creative, sustainable, corporate strategy, unifying and expertise and human-oriented qualities classify it. The concept of the "smart city" is still not limited to the use of digital, cloud-based software and services for smart cities [13, 14].

Information Communication Technologies (ICT) are used to save costs in smart cities. They optimize resource use, increase interaction and improve residents' quality of life [15, 16]. In the beginning, the focus of transport experts and smart city planners was on efficient cities road transport as well as smart development. However it shifted to "smart cities" and "smart urban mobility", symbolizing the potential of the digital age [17]. This has been proven by numerous municipal implementation projects, smart city efforts, and collaborative research funding projects emerging around the world [18], especially in the transportation sector. Due to the future mobility component, our study primarily aims to analyze the policy context prioritization for sustainable urban growth through transportation invention [18]. Knowledge is silent, it is fundamentally unexplored territory. Like any other socio-technical transformation, the move to an AI-powered mobility framework raises important questions about how the movement will be sustained [19] and which projects will represent it.

Computer driving is often considered a technique that can initiate a fundamental change in (vehicle) mobility [19, 20]. CAVs represent the pinnacle of AI in the transportation industry. Among the technologies used, machine learning and artificial intelligence may one day eliminate human contact and allow CAVs to "drive" themselves [20]. It is anticipated that CAVs can implement enhanced privacy and security standards with "ecological driving" and "energy-saving functions"; better street space allocation and trail-congestion management, as well as replacing time spent driving with time spent performing other productive tasks [21]. Despite their attention, CAVs continue to raise more unresolved concerns than concrete answers. Unmanned aerial vehicles (UAVs) and Personal aerial vehicles (PAVs) have recently emerged as a viable answer to problems especially in the fields of monitoring and evaluation and tracking surveillance [22, 23]. Fully autonomous UAVs and PAVs, logistics and human movement will be critical in the coming years; They are likely to change established travel habits in the long run by increasing urban space and vertical mass transit in cities.

Following this, the paper provides a basis for the research technique used, a description of the above-mentioned projects and a conclusion that serves as a glossary for the themes discussed in the previous sections. This part of the study also describes the findings that connect all the different pieces of the complex smart urban mobility puzzle and makes important recommendations for urban scientists, legislators, public transport, future research targets as well as city planners.

15.2 Research Method

The result of an extensive literature study examining peer-reviewed journals it comes to connecting with AI, transportation, and smart city. Although Paper One, the first publication exploring the purpose of this study was published from 1983 to 1985, the vast majority (greater than 90.5%) of the papers produced were published in the last year; several research studies were most relevant to our approach. A search highlighting minor but important aspects of research covered in the broad agenda did not find them, but it was decided that some key points needed to be supported. Discussions have been made [25, 26].

All potential combinations of key aspects of the study were explored to ensure that the majority of synergies documented in the diverse and interdisciplinary literature between AI, transportation and smart city were recognized and studied in our research. According to our search criteria, "artificial intelligence and transport" produced 750 papers, "transport and smart city" at least 550 papers and for "artificial intelligence and smart city", 180 papers were produced.

The most important research, which includes the phrases artificial intelligence (AI), transport and smart city, confirmed our purpose for writing this paper; only 18-19 papers were identified that explored all three aspects and their interrelationships. Of these studies, only two [26–28] had a direct but tangential link to the precise objectives of the present study, while the rest were more computational and "engineering-oriented" and anchored in strictly separated regions of the outlined link. A notable finding was the availability of journals and papers compared to conference proceedings; for the search terms artificial intelligence along with smart city, we briefly summarized 202 publications, with the number of research conference papers five times that [29].

Figure 15.2 illustrates the outcomes of our research for the aforementioned, along with measuring the resources discovered. Researchers have also looked for combinations of the most fundamental aspects of our research. "CAVs" and "UAVs" produced 0 papers, "CAV and PAV" 4-5 papers, "CAV and MaaS" 14-15 papers, "MaaS and UAV" 47 papers and "MaaS and "PAVs" produced 2 papers. In addition, "CAV and IoT" produced 30 papers, "UAV and IoT" 169 papers, "PAV and IoT" 3-4 papers, "MaaS and IoT" 175 papers, "PI and CAVs" produced 9 papers, "PI and UAVs" produced 29 papers.

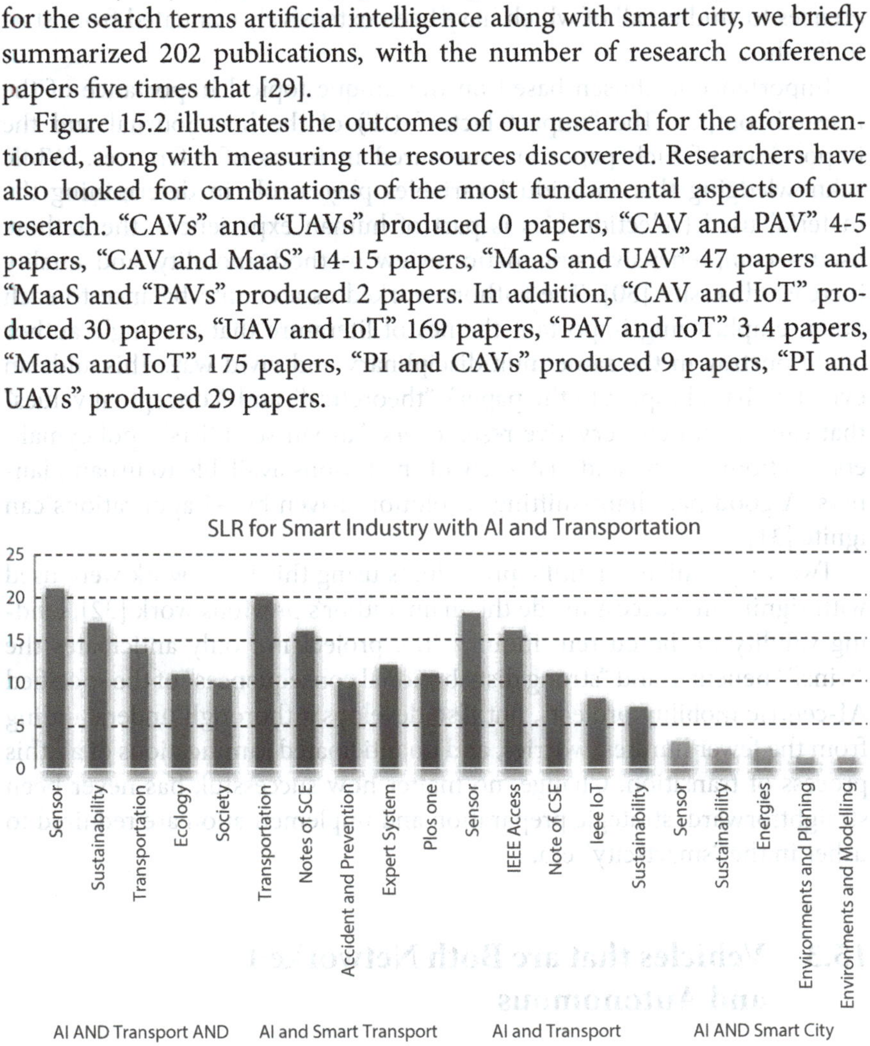

Figure 15.2 Analysis or different type research publication for Smart Industry with AI and Transportation.

In the end, each of the four authors reviewed the titles of the listed publications, a total of 1,550, Approximately 450 paper descriptions more relevant to the discussion of this research were evaluated, and an independent SLR was produced based on the paper considered to be more. The artificial intelligence that underlies next week's smart cities is important in creating the image of the transport network. All four authors then compared and synthesized their individual review versions to create a unified "broader" narrative led by the lead author. The last issue of the paper makes 135 references to studies, all of which have been extensively evaluated from start to finish.

Importance is chosen based on the unique topical importance of the research output. The "impact factor" (IF) of the host journal and the importance of each paper are measured in terms of references. While acknowledging that individual variables play a role in determining the material used (selection bias is part of human experience), the authors used a comprehensive systematic review method, mobility, and quality increase diversity [30]. The authors worked as a counterbalance to each other, emphasizing important themes of literature that no single author could portray in the same multidisciplinary and vivid way. This adds an evidence-based aspect to the paper's "theoretical" and "conceptual values" that can inform conservative researchers, "urban scientists", "policymakers", "actionmakers", and just a few of the lessons available to urban planners. A good paradigm-shifting revolution driven by AI applications can ignite [31].

Two very similar scientific procedures using this framework were used with significant success inside the main author's previous work [32], lending validity to the current inquiry. The project not only anticipates the "wins," "benefits," and "strong distributional consequences" of the specified AI-centric mobility projects, but also develops a thorough understanding from the fewer hurdles, worries, and unanticipated ramifications of all this process of transition. Change, no matter how successful, has never been straightforward; strategic preparation and implementation are required to usher in the "smart-city" era.

15.3 Vehicles that are Both Networked and Autonomous

The vehicle has altered the natural and constructed environments of the Earth more than any other technology in human history. Cities and economies are transforming [33]. Vehicles are also the organic descendants of traditional cars, with movement and activity capabilities that don't necessitate a conductor (driver) or motion-based control [34], along with interconnection features that make them aggressive, community-minded, well-informed, and organised because they can "talk" with their environment. When IoT is in place, it will eventually be used by each one and each thing. CAVs will transform mobility, transport links, and transportation systems by shifting control of the vehicle and operating tasks away from humans and toward computers endowed with huge AI and wireless interconnecting capabilities. CAVs are predicted to become another global standard in transport, transforming cars and city social progress by the use of AI.

Because CAVs are significantly different from traditional human-driven vehicles, the auto industry will need to completely overhaul and modernize its production methods. 1. Audi, 2. Baidu, 3. BMW, 4. Daimler, 5. Delphi, 6. DidiChuxing, 7. Ford, 8. GM, 9. Honda, 10. Huawei, 11. Hyundai, 12. Jaguar Land -Rover, 13. Lyft, 14. Magna, 15. Mercedes-Bosch alliance, 16. Microsoft, autonomous, 17. PSA, 18. Renault-Nissan alliance, 19. Samsung, 20. Tesla, 21. Toyota, 22. Ubisoft and others are currently developing CAVs or conducting trials with varying degrees of effectiveness. Giants that provide intelligence like search engines and giants that travel like Uber, which are not typically considered typical vehicle manufacturers, will significantly change the dynamic behaviour of the company and become new competitors in a stronger and more flexible automotive market. ICT is trying to distinguish strategic advantage [34]. If smart cities decide to massively promote self-sustaining growth by reducing motor vehicle ownership, the industrial sector may try to force new simple values to prioritize sharing use over the investment portfolio.

The United States (USA), the United Kingdom, Australia, New Zealand (NZ), Germany (DE), Sweden and China are major countries that have positioned themselves at the geographic centre of the CAV "explosion" by enforcing the legislation that enables CAV and investing billions in R&D projects such as trials and trials with CAVs [35]. Every rebellion against the norm continues to promise optimism and great improvements. CAVs are expected to provide numerous benefits, including increased leisure time (for a typical driver this equates to six weeks each year); increased rail safety and disaster avoidance; in web applications, convenience or driving pleasure; perhaps by simplifying the judgment call to prioritize ridesharing and ridesharing marketing tactics, reducing road congestion, air degradation, poor air quality, traffic noise and other pollutants [36].

At the same time, even in a transition period when CAVs coexist with simple cars, a few concerns are generally less interactable and communicated and researched about increased manipulation, traffic and emissions to the atmosphere. There are difficulties with malware, s/w and s/w challenges, privacy losses and manipulation of travel data, distribution of responsibility CAVs, semi-autonomous and CAVs. These opportunities and challenges are documented in the literature [17–20], but rarely in a single source. Table 15.1 summarizes the primary possibilities; reflective restrictions. Based on these observations it will be used for the future, carrying out the full-scale launch of the CAV [37].

Despite technical advances, a host of promised benefits, and millions of kilometres travelled autonomously in segregated and regulated environments around the world, CAVs may not yet be easily incorporated into our urban transport networks. This is a time-consuming technique that may take years to complete [24]. The emergence of these cars is followed by uncertainties about the vehicle industry and their impact on land use practices, but more importantly, how and why the public formulates decisions about benefits and concerns [37]. The spontaneous, computerized deployment of CAVs may conflict with environmental and societal objectives, ignoring the fact that the public transport system is a complex social-economic system that requires collaborative optimization of both technological and social subsystems [38].

As technology is only one of the myriad tools in a mobility toolkit [19, 31, 36], the transformation of the CAV must be supported by the regulatory, moral, academic, commercial and socialization framework that specifies a particular mode of operation. Because of their traditionally conflicting roles, merging cars and public transport, even in their subsequent upgrades to mechanized, can be an unhappy union, but future policy and planning initiatives should help in the complementary tensions and achieve an ideal blend between the two [39]. CAVs should be more flawless than they are.

Choosing to be electric is a possibility, but only if the power is generated from renewable energy sources and therefore does not pollute the environment. However, the path to transformation is not simple. To assist the transition, the following improvements for CAV policies and strategy should be addressed: advanced technologies; legislative action; recession and job opportunities; ethics; road systems and land use; forced integration; trace security; cyber security and privacy; promotional events; follow the crowd; and travel times; and ultimately

Table 15.1 Possibilities and difficulties in the development, adoption, and utilisation of Connected and Autonomous-Vehicles (CAVs).

S. N.	**Potential societal and user benefits**	**Potential issues for society and users**
PD1	Decrease the personal mistake aspect in driving, resulting in increased trace safety and accident prevention.	Ambiguity in terms of accountability in accident and damage scenarios.
PD2	Enhanced trace security as a result of increased and simplified monitoring, tracking, and monitoring.	heightened susceptibility to S/W and H/W defects, as well as cyber-security risks.
PD3	Reduced traffic congestion as a result of more efficient mobility and parking management.	More users and empty automobiles may result in additional car journeys.
PD4	Efficiency gains as a result of effective routing; platooning and steady track flows.	Issues connecting with non- and partially automatic vehicles, among other things.
PD5	Noise pollution will be reduced from the engines of CAVs and they will be quieter.	In new technology and the authorities in charge of administering CAVs.
PD6	CAVs will eco-drive, reducing energy use and reliance on fossil fuels.	Concerns about confidentiality and a lack of confidentiality space.
PD7	Increased productivity–people can utilize their in-car time to perform useful things.	Threats to employability - driving-related employment will cease to exist, necessitating reskilling labour.

(Continued)

Table 15.1 Possibilities and difficulties in the development, adoption, and utilisation of Connected and Autonomous-Vehicles (CAVs). (*Continued*)

S. N.	Potential societal and user benefits	Potential issues for society and users
PD8	There is a lot of potential for vehicle sharing and transportation sharing.	People prefer or are accustomed to privately owned automobiles; therefore "ownership" rights are likely to be lost.
PD9	Substantial-demand responsiveness.	Possible setback for public transportation as we know it.
PD10	Cost reduction for logistics; taxi; and ride-sharing firms; as well as their clients.	Investments in modern roadinfrastructure fit the demands of CAVs at a high cost.
PD11	There will be less demand for parking, which will free up public space for more people-oriented purposes.	Rural inequities if CAVs become a cost-cutting measure.
PD12	There are fewer levels of social marginalization - fewer hurdles to "driving" a vehicle due to age, disability, and expertise.	Inequity concerns may arise if CAVs become overly expensive and intricate privately owned computers.
PD13	Reduced standards for enforcing and policing.	There is a need for a completely new road transportation rules framework and track code of practice.
PD14	Nicer rides as a result of less acceleration and deceleration, as well as faster speeds.	Problems with behaviour adaption - change takes time and causes displeasure.
PD15	There is more interior room because there is no need for a steering wheel.	Driving abilities and situational awareness are deteriorating.

relevance, confidence, and customer enthusiasm. Even if there are suitable solutions for all of the important difficulties listed above, the vehicle fleet of the future will not be nearly homogeneous before 2051, and only if CAV prices stay cheap. The path to transformation, however, won't be difficult. According to [39], the following issues are important to the planning and management of CAVs must be provided for change to proceed smoothly: technology, legislation; professional ethics and crises; road systems and land use, assimilation; road safety, cyber security and privacy; marketing strategies; road congestion; travel behaviour; and finally, availability, trust and customer readiness. If all of the critical challenges outlined above are not addressed before 2050, or if the costs of CAVs are relatively low, the future vehicle fleet is likely to be homogeneous [40].

15.4 Personal Aerial Automobile Vehicles and Unmanned Aerial Automobile Vehicles

Drones or unmanned aerial vehicle systems are powerful new interventions in aviation and transportation based on wireless technology and artificial intelligence. In the future, unmanned aerial vehicles can provide robust solutions for advanced military, law enforcement, public safety, large batteries, long-range and low-power micro-radio devices and low-cost airframes. Powerful microprocessors and engines will be available, law enforcement applications, environmental monitoring, surveillance and threat detection [40]. UAVs can assist with intelligence; surveillance and discovery; border patrol; goal identification and setting; counter-insurgency operations; attack and strike; cyber security applications; law enforcement; and environmental monitoring; and geographic activity; satellite images; aerial mapping; weather forecasting and meteorological; wildfire warning; traffic control; freight transport; accident records; emergency search and rescue; and other responsibilities. Many of these occupations act as barometers for the daily functionality of a smart city. Before air technologies are fully integrated into the smart city environment, barriers such as the absence of an aircraft pilot to monitor and avoid other aircraft, as well as a wide variety of drone missions or capabilities, must be addressed [41].

Combining unmanned aerial vehicles (UAVs) with smart cities solves problems in a more responsible economic environment. Business concerns such as morality and privacy, expense, licenses, legal action and market compliance, as well as a relaxing lifestyle if there are technical problems such as correct use of wireless sensors, data transmission, system management, asset consulting and allocation and power control are completely resolved. Manned Aerial Vehicles (MAVs), Personalized Aircraft (PAVs) and Personal-Air-Transport Systems (PATS) create forms of mobility that claim to combine the benefits of ground level and "air transport" to minimize congestion problems using "free space". In aircraft, PAVs are a unique mobility solution that will bridge the gap between scheduled aircraft and local transport by enabling historic speed levels; and optional urban transport [47].

If improved automation and reduced cost are realized, PAVs will eventually be able to serve urbanized areas of smart cities. PAVs will be able to carry passengers without the need for flight capability and will become more practical for each. Transport infrastructure includes the amount of landing and parking spaces, convenient flight routes, cooperation with other modalities, and synergy with autonomous drones. If properly planned and constructed, PAVs can greatly reduce the occurrence of increased congestion. To date, we have published approximately 32K (Scopus indexed conference and journal research) paper findings that are almost entirely related to the technology and engineering aspects of the two findings. Both are expected to be fundamental game-changers in urban transport and facilitators of the idea of a smart city in the technology of the future, according to previous research [42].

15.5 Mobile Connectivity as a Service

A powerful idea, still in its early stages of development in terms of full-scale implementation, but highly overrated, Mobility as a Service (MaaS) offers special digital intermodal transport packages that private cars have. It will be phased out as a result of a full-featured smart online website capable of providing comprehensive location-based services, scheduling, smart pricing and real information solutions.

MaaS has the potential to drastically reduce the number of vehicles on the roads and result in travel delays if it becomes a real opposition to personal vehicles, making automobile use a necessary activity. It has resulted in significant reductions in transport, air pollution, noise disturbance, energy use and social exclusion, as well as gains in road efficiency and safety education, healthy lifestyles, social cohesion and community stability [43]. The specifics of MaaS implementation make the term vague and fully integrated, and despite the lack of a uniform definition, these data show that MaaS research will never be slowed by demand in a short time. There is now a split between researchers and developers who see MaaS as just a driving and car-sharing scheme enabler and others who think MaaS should be even more. This makes it difficult to assess whether MaaS can change travel behaviour and help build the world, as MaaS has not yet been fully characterized or contextualized [44]. There is really limited perspective to date in terms of the behavioural impact of MaaS or its future potential, particularly noteworthy in this area.

15.6 Major Role for Smart City Development with IoT and Industry 4.0

Whereas the term Internet of Things (IoT) is growing more popular, there is no accepted definition or understanding of what IoT entails [46]. For example, the IoT began as a network environment that envisioned a foreseeable future where everyday products would be loaded with sensors and appropriate protocol stacks with embedded systems and electronic transmitters, allowing them to connect with and allow users. Becoming an important component of the Internet [46], it uses Internet-of-Things, integrated cloud architecture of networks, S/W-developer, sensor technology, human interfaces, Generation 5 and "data analytics" to generate value. The IoT is the technological backbone and enabler of smart cities, since urban areas require three critical aspects that IoT can provide: intelligence, connectivity and measurement [47].

Each of these numbers reflects a significant drop in sensor or IoT component pricing, making it commercially viable to install a sensor behind both key parts and connect it to the Internet. Predictive Maintenance is an example where spare parts of investment products (for example, large machines) are monitored by IoT sensors and when a certain level of wear and tear is reached, the need is returned to logistics. Monitoring application [46, 47] this allows us to reduce (preferably eliminate) on-site spare parts storage while maintaining a near-ideal level of service for machine spare parts.

Based on the idea of an object-object relationship, IoT technology [45] can provide the hardware (h/w) and software (s/w) elements of IoT using such gadgets.

Radio-frequency identification dockets (RFID) or readers, sensor technologies, positioning systems, evolving technologies to collect data on road

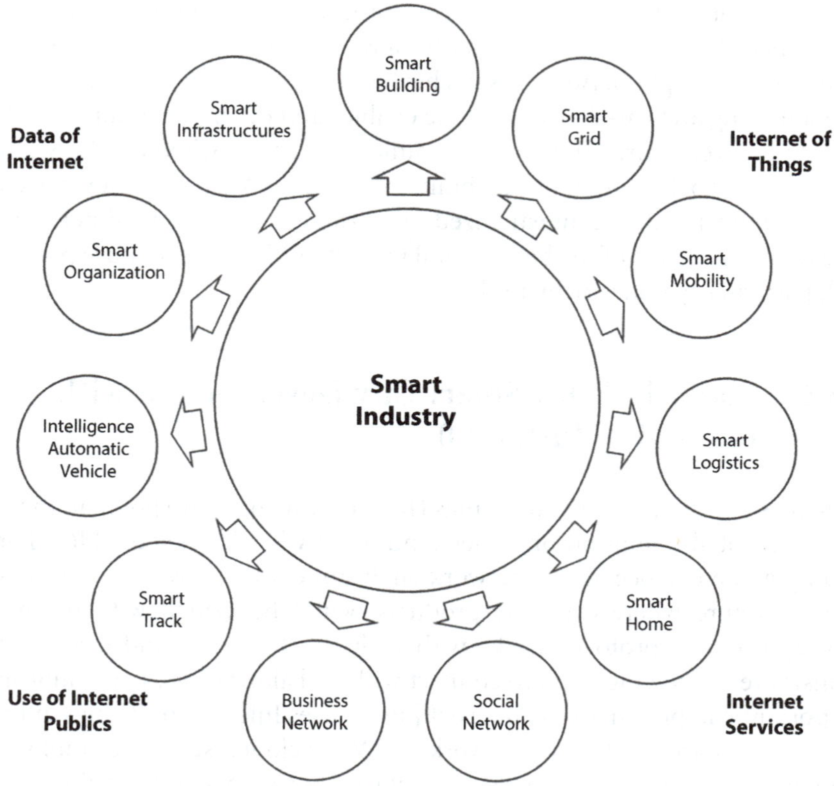

Figure 15.3 Smart industry role for internet technologies.

conditions, traffic accidents, road repairs and their whereabouts; structuring will take place in the ecosystem [47–51]. "Physical internet and industry-4.0" will be two additional important facilitators of artificial intelligence-transport, "smart city" connectivity. PI is a new idea for freight transport and logistics aimed at improving the environment; carried out; social efficiency; hidden; provided; sustainability of how physical things are transported; economic; and used internationally [52]. PI is a unique concept for freight logistics and distribution aimed at improving the ecological; social efficiency and economics; and the sustainability of how physical items are transported, recorded, produced, supplied and used around the world [53]. Creatively and effectively arranging the critical components of a logistics system will assist in the pursuit of a smart, green city. PI is the transport and logistics industry's response to "Industry 4.0" [54]. Figure 15.3 represents Application of Smart industry and role for internet technologies for autonomous industries.

Often referred to as the 4th industrial revolution with industry 4.0, it emerged as a similar strategic effort for industrial innovation and expanded into the extraordinary changes that typified computerized systems, computerization, digital and industrial data system; Cloud computing, Cyber-Physical-Systems (CPSs), Enterprise Resource Planning (ERP) and Internet of Things (IoT) [53–55]. Technologies IoT, PI and Industry 4.0 are extensively covered in the bibliography, especially in terms of technological and systems engineering; outputs of approximately 53K, 10.5K and 26K. Scopus indexing conference and journal research papers are currently available on an online platform [56]. However, it is very important to properly analyze the latest status of previous programs in terms of smart city developments.

15.7 Conclusion

This paper focuses on an attempt to create a comprehensive and systematic review for all elements of the future technology of the AI-based smart city transport ecosystem, after contextualizing and connecting them is a wide range of urban science professionals interested in understanding. This will hopefully increase the visibility and usability of what is still unstudied, at least in the whole of AI-transport smart city connectivity. The current initiative recognizes the transformative capability of AI in the context of technology and its potential to be a peer force to transform mobility in new and unexpected ways, a comprehensive user-centred paradigm that understands and delights the individual customer, market or society in general. The promise of self-driving, connected, shared and digitized transport services are insufficient unless it supports efforts that contribute to environmental protection and resource efficiency, productivity gains, social inclusion, integration, health and well-being. Citizens should consider that change can benefit many individuals, that they can become active and engaged participants in the new urban environment, particularly taking advantage of the potential provided by the artificial intelligence-transport-based potential. The key to achieving this smart city connection, as with any cultural exchange, is living in harmony. With the high-tech transformation happening and maximizing its likelihood, there could still be risks and harmful side effects, due to a flawed MaaS business strategy, such as the potential increase in vehicle tracks due to the emergence of fuller and (scary) deserted CAV rides. This should be avoided. Intelligent innovation must also include important socio-centric components. In addition to making significant changes in an age characterized by living, traditionally empowered, privately held, disconnected and ground-care, its enormous impact on urban growth that controls how cities are truly planned and control a centuries-old structure is a thoughtful focus on tenacity, cohesion and agility. It requires a plan. Innovations in transport procurement cannot be accelerated unless the relevant technical, regulatory, pedagogical and ethical foundations are strongly in place and mature enough to withstand unexpected and unexpected challenges.

References

1. Li, L.; Lin, Y.L.; Zheng, N.N.; Wang, F.Y.; Liu, Y.; Cao, D.; Wang, K.; Huang, W.L. Artificial intelligence test: A case study of intelligent vehicles. *Artif. Intell. Rev.* 2018, 50, 441–465.
2. Abduljabbar, R.; Dia, H.; Liyanage, S.; Bagloee, S.A. Applications of artificial intelligence in transport: An overview. *Sustainability* 2019, 11, 189.
3. Knowles, R.D.; Ferbrache, F.; Nikitas, A. Transport's historical, contemporary and future role in shaping urban development: Re-evaluating transit-oriented development. *Cities* 2020, 99, 102607.
4. Giyenko, A.; Im Cho, Y. Intelligent UAV in smart cities using IoT. In *Proceedings of the 2016 16th International Conference on Control, Automation and Systems (ICCAS), Gyeongju, Korea,* 16–19 October 2016; pp. 207–210.
5. Lyons, G. Getting smart about urban mobility–aligning the paradigms of smart and sustainable. *Transp. Res. Part A Policy Pract.* 2018, 115, 4–14.
6. Kaplan, A.; Haenlein, M. Siri, Siri, in my hand: Who's the fairest in the land? On the interpretations, illustrations, and implications of artificial intelligence. *Bus. Horiz.* 2019, 62, 15–25.
7. Gurkaynak, G.; Yilmaz, I.; Haksever, G. Stifling artificial intelligence: Human perils. *Comput. Law Secur. Rev.* 2016, 32, 749–758.
8. Everitt, T.; Hutter, M. Universal artificial intelligence. In *Foundations of Trusted Autonomy*; Springer: Cham, Switzerland, 2018; pp. 15–46.
9. Ahvenniemi, H.; Huovila, A.; Pinto-Seppä, I.; Airaksinen, M. What are the differences between sustainable and smart cities? *Cities* 2017, 60, 234–245.
10. Docherty, I.; Marsden, G.; Anable, J. The governance of smart mobility. *Transp. Res. Part A Policy Pract.* 2018, 115, 114–125.
11. Milakis, D.; Van Arem, B.; Van Wee, B. Policy and society related implications of automated driving: A review of literature and directions for future research. *J. Intell. Transp. Syst.* 2017, 21, 324–348.
12. Papa, E.; Ferreira, A. Sustainable accessibility and the implementation of automated vehicles: Identifying critical decisions. Urban Sci. 2018, 2, 5.
13. Thomopoulos, N.; Nikitas, A. Smart Urban Mobility Futures: Editorial for Special Issue. *Int. J. Automot. Technol. Manag.* 2019, 19, 1–9.

14. Nikitas, A.; TchouamouNjoya, E.; Dani, S. Examining the myths of connected and autonomous vehicles: Analysing the pathway to a driverless mobility paradigm. *Int. J. Automot. Technol. Manag.* 2019, 19, 10–30.
15. Nikitas, A.; Kougias, I.; Alyavina, E.; NjoyaTchouamou, E. How can autonomous and connected vehicles, electromobility, BRT, hyperloop, shared use mobility and mobility-as-a-service shape transport futures for the context of smart cities? *Urban Sci.* 2017, 1, 36.
16. Barmpounakis, E.N.; Vlahogianni, E.I.; Golias, J.C. Unmanned Aerial Aircraft Systems for transportation engineering: Current practice and future challenges. *Int. J. Transp. Sci. Technol.* 2016, 5, 111–122.
17. Bakogiannis, E.; Kyriakidis, C.; Zafeiris, V. Using Unmanned Aerial Vehicles (UAVs) to analyze the urban environment. *Eur. J. Eng. Form. Sci.* 2019, 3, 20–28.
18. Hofmann, E.; Rüsch, M. Industry 4.0 and the current status as well as future prospects on logistics. *Comput. Ind.* 2017, 89, 23–34.
19. Ingwersen, P.; Serrano-López, A.E. Smart city research 1990–2016. *Scientometrics* 2018, 117, 1205–1236.
20. Zear, A.; Singh, P.K.; Singh, Y. Intelligent transport system: A progressive review. *Indian J. Sci. Technol.* 2016, 9, 1–8.
21. Gandia, R.; Antonialli, F.; Cavazza, B.H.; Neto, A.M.; De Lima, D.A.; Sugano, J.Y.; Nicolai, I.; Zambalde, A.L. Autonomous vehicles: Scientometric and bibliometric review. *Transp. Rev.* 2019, 39, 9–28.
22. Chehri, A.; Mouftah, H.T. Autonomous vehicles in the sustainable cities, the beginning of a green adventure. *Sustain. Cities Soc.* 2019, 51, 101751.
23. Gavanas, N. Autonomous Road Vehicles: Challenges for Urban Planning in European Cities. *Urban Sci.* 2019, 3, 61.
24. Rawat, R., Rajawat, A. S., Mahor, V., Shaw, R. N., & Ghosh, A. (2021). Surveillance Robot in Cyber Intelligence for Vulnerability Detection. In *Machine Learning for Robotics Applications* (pp. 107–123). Springer, Singapore.
25. Rawat, R., Mahor, V., Chirgaiya, S., & Rathore, A. S. (2021). Applications of Social Network Analysis to Managing the Investigation of Suspicious Activities in Social Media Platforms. In *Advances in Cybersecurity Management* (pp. 315-335). Springer, Cham.
26. Rawat, R., Mahor, V., Rawat, A., Garg, B., & Telang, S. (2021). Digital Transformation of Cyber Crime for Chip-Enabled Hacking. In *Handbook of Research on Advancing Cybersecurity for Digital Transformation* (pp. 227-243). IGI Global.
27. Rawat, R., Rajawat, A. S., Mahor, V., Shaw, R. N., & Ghosh, A. (2021). Dark Web— Onion Hidden Service Discovery and Crawling for Profiling Morphing, Unstructured Crime and Vulnerabilities Prediction. In *Innovations in Electrical and Electronic Engineering* (pp. 717-734). Springer, Singapore.
28. De Almeida Correia, G.H.; Milakis, D.; van Arem, B.; Hoogendoorn, R. Vehicle automation and transport system performance. In *Handbook*

on Transport and Urban Planning in the Developed World; Edward Elgar Publishing: Cheltenham, UK, 2016; pp. 498–516.

29. Bansal, P.; Kockelman, K.M. Are we ready to embrace connected and self-driving vehicles? A case study of Texans. *Transportation* 2018, 45, 641–675.
30. Milakis, D. Long-term implications of automated vehicles: An introduction. Transp. Rev. 2019, 39, 1–8.
31. Sochor, J.; Nikitas, A. Vulnerable users' perceptions of transport technologies. *Proceedings of the Institution of Civil Engineers. Urban Des. Plan.* 2016, 169, 154–162.
32. Menouar, H.; Guvenc, I.; Akkaya, K.; Uluagac, A.S.; Kadri, A.; Tuncer, A. UAV-enabled intelligent transportation systems for the smart city: Applications and challenges. *IEEE Commun. Mag.* 2017, 55, 22–28.
33. Mohamed, N.; Al-Jaroodi, J.; Jawhar, I.; Idries, A.; Mohammed, F. Unmanned aerial vehicles applications in future smart cities. *Technol. Forecast. Soc. Chang.* 2018, 153, 119293.
34. Balac, M.; Vetrella, A.R.; Rothfeld, R.; Schmid, B. Demand estimation for aerial vehicles in urban settings. *IEEE Intell. Transp. Syst. Mag.* 2019, 11, 105–116.
35. Liu, Y.; Kreimeier, M.; Stumpf, E.; Zhou, Y.; Liu, H. Overview of recent endeavors on personal aerial vehicles: A focus on the US and Europe led research activities. *Prog. Aerosp. Sci.* 2017, 91, 53–66.
36. Lichfield, G. When Will We Have Flying Cars? Maybe Sooner Than You Think. *MIT Technology Review*. 2019. Available online: https://www.technologyreview.com/s/612891/when-will-we-have-flying-cars-maybe-sooner-than-you-think/ (accessed on 5 March 2020).
37. Alonso-González, M.J.; Hoogendoorn-Lanser, S.; van Oort, N.; Cats, O.; Hoogendoorn, S. Drivers and barriers in adopting Mobility as a Service (MaaS)–A latent class cluster analysis of attitudes. *Transp. Res. Part A Policy Pract.* 2020, 132, 378–401.
38. Hirschhorn, F.; Paulsson, A.; Sørensen, C.H.; Veeneman, W. Public transport regimes and mobility as a service: Governance approaches in Amsterdam, Birmingham, and Helsinki. *Transp. Res. Part A Policy Pract.* 2019, 130, 178–191.
39. Smith, G.; Sochor, J.; Karlsson, I.M. Mobility as a Service: Development scenarios and implications for public transport. *Res. Transp. Econ.* 2018, 69, 592–599.
40. Hensher, D.A. Future bus transport contracts under a mobility as a service (MaaS) regime in the digital age: Are they likely to change? *Transp. Res. Part A Policy Pract.* 2017, 98, 86–96.
41. Polydoropoulou, A.; Pagoni, I.; Tsirimpa, A.; Roumboutsos, A.; Kamargianni, M.; Tsouros, I. Prototype business models for Mobility-as-a-Service. *Transp. Res. Part A Policy Pract.* 2020, 131, 149–162.
42. Matyas, M.; Kamargianni, M. The potential of mobility as a service bundles as a mobility management tool. *Transportation* 2019, 46, 1951–1968.

43. Xie, Y.; Danaf, M.; Azevedo, C.L.; Akkinepally, A.P.; Atasoy, B.; Jeong, K.; Seshadri, R.; Ben-Akiva, M. Behavioral modeling of on-demand mobility services: General framework and application to sustainable travel incentives. *Transportation* 2019, 46, 2017–2039.
44. Schikofsky, J.; Dannewald, T.; Kowald, M. Exploring motivational mechanisms behind the intention to adopt mobility as a service (MaaS): Insights from Germany. *Transp. Res. Part A Policy Pract.* 2020, 131, 296–312.
45. Pangbourne, K.; Mladenovic′, M.N.; Stead, D.; Milakis, D. Questioning Mobility as a Service: Unanticipated implications for society and governance. *Transp. Res. Part A Policy Pract.* 2020, 131, 35–49.
46. Chang, S.J.; Chen, H.Y.; Chen, H.C. Mobility as a service policy planning, deployments and trials in Taiwan. *Iatss Res.* 2019, 43, 210–218.
47. Mulley, C.; Ho, C.; Balbontin, C.; Hensher, D.; Stevens, L.; Nelson, J.D.; Wright, S. Mobility as a service in community transport in Australia: Can it provide a sustainable future? *Transp. Res. Part A Policy Pract.* 2020, 131, 107–122.
48. Arias-Molinares, D.; García-Palomares, J.C. The Ws of MaaS: Understanding mobility as a service from a literature review. *IATSS Res.* 2020, in press.
49. Bothos, E.; Magoutas, B.; Arnaoutaki, K.; Mentzas, G. Leveraging Blockchain for Open Mobility-as-a-Service Ecosystems. In *Proceedings of the IEEE/WIC/ACM International Conference on Web Intelligence-Companion Volume, Thessaloniki, Greece,* 14–17 October 2019; pp. 292–296.
50. Stocker, A.; Shaheen, S. Shared Automated Mobility: Early Exploration and Potential Impacts. *Road Veh. Autom.* 2018, 4, 125–139.
51. Mulley, C. Mobility as a Services (MaaS)—Does it have critical mass? Transp. Rev. 2017, 37, 247–251.
52. Jittrapirom, P.; Marchau, V.; van der Heijden, R.; Meurs, H. Future implementation of Mobility as a Service (MaaS): Results of an international Delphi study. *Travel Behav. Soc.* 2018, in press.
53. Xu, L.D.; Xu, E.L.; Li, L. Industry 4.0: State of the art and future trends. *Int. J. Prod. Res.* 2018, 56, 2941–2962.
54. Rawat, R., Mahor, V., Chirgaiya, S., & Garg, B. (2021). Artificial Cyber Espionage Based Protection of Technological Enabled Automated Cities Infrastructure by Dark Web Cyber Offender. In *Intelligence of Things: AI-IoT Based Critical-Applications and Innovations* (pp. 167–188). Springer, Cham.
55. Rawat, R., Garg, B., Mahor, V., Chouhan, M., Pachlasiya, K., &Telang, S. Cyber Threat Exploitation and Growth during COVID-19 Times. In *Advanced Smart Computing Technologies in Cybersecurity and Forensics* (pp. 85–101). CRC Press.
56. Mahor, V., Rawat, R., Kumar, A., Chouhan, M., Shaw, R. N., & Ghosh, A. (2021, September). Cyber Warfare Threat Categorization on CPS by Dark Web Terrorist. In *2021 IEEE 4th International Conference on Computing, Power and Communication Technologies (GUCON)* (pp. 1–6). IEEE.

Index

Printed in the USA/Agawam, MA
September 16, 2026

910661.078